NOV 2 8 2010

NO
SEA

D1008844

A SHORT HISTORY OF THE
Honey Bee

A SHORT HISTORY OF THE
Honey Bee

*Humans, Flowers, and Bees in the
Eternal Chase for Honey*

Images by
ILONA

Text by
E. READICKER-HENDERSON

TIMBER PRESS
PORTLAND • LONDON

Images copyright © 2009 by Ilona McCarty.
Text copyright © 2009 by E. Readicker-Henderson.
All rights reserved.

Published in 2009 by Timber Press, Inc.

The Haseltine Building   2 The Quadrant
133 S.W. Second Avenue, Suite 450   135 Salusbury Road
Portland, Oregon 97204-3527   London NW6 6RJ
www.timberpress.com   www.timberpress.co.uk

Printed in China

Library of Congress Cataloging-in-Publication Data

Ilona.
   A short history of the honey bee : humans, flowers, and bees in the eternal
chase for honey / images by Ilona ; text by E. Readicker-Henderson.
      p. cm.
   Includes bibliographical references.
   ISBN 978-0-88192-942-3
   1. Honeybee—Pictorial works. 2. Honeybee. 3. Bee culture—Pictorial
works. 4. Bee culture. 5. Honey. I. Readicker-Henderson, Ed. II. Title.
   SF523.7.I45 2009
   638'.1—dc22
                                                        2008050286

A catalog record for this book is also available from the British Library.

CONTENTS

Acknowledgments

THANKS TO THE MANY people who generously lent their time and expertise, as well as the opportunity to photograph their hives, bees, and honey, to help make this book a reality: Bill Lewis, Bill's Bees; Lisa Blackburn, The Huntington Library, Art Collections, and Botanical Gardens; Ted Dennard, The Savannah Bee Company; Ruth Ingles, Honey House; Charles Kennard; Gina Knudson; Andrea Langworthy, The Naked Bee; J. Douglas McGinnis, Tropical Blossom Honey Company; Dan and Lisa Mudd, Salmon Valley Honey; Randy, Eric, and Ian Oliver; Gus Rouse, Kona Queen Hawaii Inc.; and Richard Spiegel, Volcano Island Honey Company. Without their help and cooperation this book would not have been possible. Thanks to E. Readicker-Henderson for the poetry of his words.

— ILONA

My parents, who somehow thought bringing beehives into the suburbs was sensible; Leslie, who showed where the words started; M, who opened the first flower of the story; IG, who saw the whole field; Cleo, for providing shelter from the storm; and SS, who showed there was no storm at all.

— E. READICKER-HENDERSON

We would like to thank Helen Townes and Tom Fischer, who saw what this book could be, and helped make it the fine thing you have in your hands.

ONE:
The First Taste

MY FATHER HAS a dangerous combination of a straight face and a strange sense of humor, so it was no wonder that none of us believed him when he said he was going out to buy bees.

We lived in the suburbs. People in the 'burbs don't keep bees.

But that night, after the moonlit air had the heavy scent of orange blossoms that only comes for a few weeks, Dad drove up, and in the back of his International Harvester pickup truck—its air conditioner so effective it spit ice—was a single hive of bees.

A white box, maybe three feet tall, two feet to each side.

This did not, let us admit right up front, go over well. My mother fretted about complaining neighbors, my sisters objected to the very idea of bringing bugs to the house. My brother was all for it, but that only lasted until the moment he poked his face into the hive and discovered he was desperately allergic to bee stings.

Even the dog was suspicious.

So while my father and a friend moved the hive into the backyard, I took what was surely my last look at the place where my friends and I had always played catch.

How could we ever do it again? The instant we stepped out the door, surely we'd be swarmed, attacked. As soon as those bees were set down, our yard would be a no-man's-land. A no-kid's-land.

And yet.

Raw honeycomb, the purest way to enjoy honey.

The next morning, when I woke and peered hesitantly outside through the back windows, afraid I'd be able to feel the vibration of bee buzz even through the glass, it was immediately clear the bees had their own agenda. Their own work.

As soon as the sun had come up, the bees had barely paused to notice their change of address before packing off to the job. In fact, by the time I peeked over the sill, they were already well into their day, entirely busy, busy as bees, going about their fine business of transforming the world into a taste, gathering up the terrain, flower by flower.

It would be a while before I understood this, but the bees were simply doing what bees do: acting as the gardeners of the world and making their incredibly generous gift of the landscape.

Looking for a reason to be hopeful, I checked my *Boy Scout Handbook*. Yes, there was even a beekeeping merit badge. Maybe this wouldn't be so bad, after all.

Just how good it could be, though, wasn't revealed for a while, not until we'd had the bees for a few months, not until they'd simply become another fact of home, like the stiff, dry grass underfoot. Not until after even my most skittish friends realized the bees weren't going to interrupt our backyard games of catch.

One fall morning, my father went out dressed in full bee regalia—the white hood with the net over his face, the high, stiff yellow rubber gloves, his pants tucked down into his boots.

And that day, I finally learned the true magic of bees. More: I found my quest.

Dad lit a match to some newspapers, stuffed them into the bellows, which he squeezed at the hive entrance and around the openings of the top, pouring smoke onto the bees. Then he pried open the lid of the hive, took a frame of heavy comb out, holding it yel-

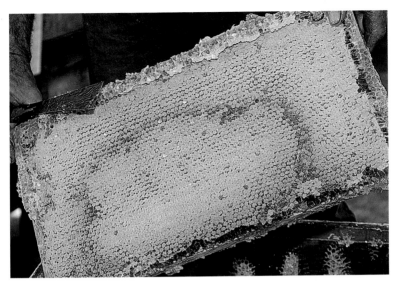

A frame, fresh from the hive, heavy with honey.

low and rich in the light. He drew his pocketknife across the wax caps, and the honey began to flow, an amber that made me think of what I'd heard in science class, that it was possible to slow light down so much that it became solid.

Dad held the comb out, and my world irrevocably changed. Changed in a single, hesitant touch, tongue to honey. The pure taste of sweetness.

Nearly forty years later, I'm still chasing that taste of honey straight out of the hive, honey that I saw made bee by bee, flower by flower, before we spun it out under the centrifugal force of the extractor and our whole house smelled like the inside of the world's most delicious honey jar.

Forget the wine snobs who tell you that what they drink is the essence of the country. Honey is more than that: it is the truest

OVERLEAF: *Frame of honeycomb ready to go into the extractor. The wax caps have been cut from the cells, and the honey is just beginning to drip out.*

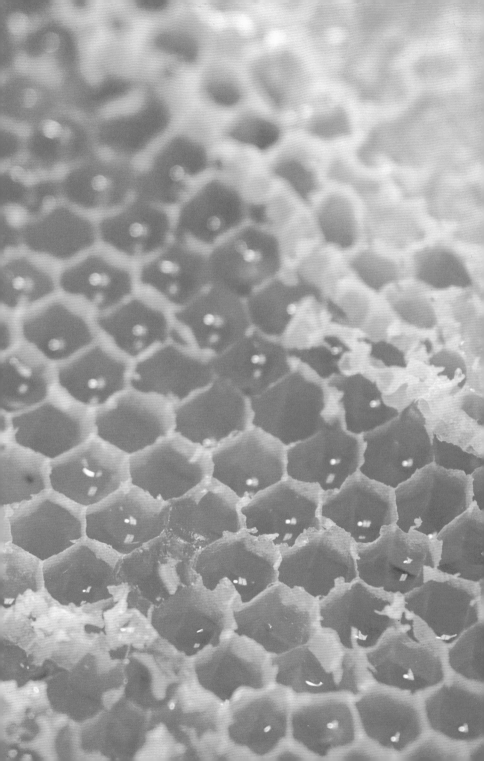

distillation of the landscape, as specific to a place as the way sunlight hits flowers in the morning.

We had lived in that suburban house for two or three years by the time we were joined by the bees, but when I tasted my first drop of true local honey, I realized that up to that instant, I had known nothing whatsoever about my own home. Suddenly, my senses were filled with the globemallows, the undercurrent of ironwood, the cool air of a breeze coming off brittlebush in summer. Our backyard honey held the lightning that blinded night in autumn and drew the creosote scent from the rocks, the dust storms that rolled in ahead of the rain.

In that taste was a revelation. What usually gets passed off as honey, sold in grocery stores, put on diner tables in a plastic bear, is a blend, from no clear source or geography, maybe not even all from the same country or time zone. And that's a generic sweetness, pleasant but muddied, like a child's finger-painting that has gone on too long, the colors becoming indistinct.

Pure honey, though, honey treasured and nourished and definitive of its own place and time, honey drawn from a single hive or hives in a single field, offers a lesson in the vital difference between flowers that grow in sand or in loam, those that get sun in the morning versus the ones that see only a little light in the afternoon.

That kind of honey is an element, as clean as oxygen.

Although it all goes by the same name, honey that started off as, say, orange blossoms, is nothing at all like honey made from the nectar of wildflowers, or from the fragile blue of desert lavender or just-after-dawn gold of catsclaw. Even honey from the same species of plant can look and taste vastly different—clover honey

Yellow sweet clover, alfalfa, alsike clover, and sainfoin: some of the bees' favorite foods.

ranges from pale to a gold so deep that it seems like it's trying to hide the secrets of its miraculous flavor.

Even within that single species of flower, there is a gradation of taste—from the delicate taste of light to the strong notes of dark.

And if you change only one tiny aspect of true honey—add a single blossom from a different type of flower—it's like making the world new again from scratch, hanging stars in the sky in patterns that rearrange to no known constellation. A dangerous alchemy that can pay off like a perfume tuned to the neck of a beautiful woman, or can simply fail and disappear into the bland.

A jar of honey, I discovered, is a voyage of discovery, a promise of risk and reward.

The honey I've brought home from the deserts of the Middle East tastes nothing like what has come back with me from the jungles of South America, from the clean plains of the steppe. Heather honey from Scotland is strong enough to cut into the flavors of a land where they eat kippers for breakfast; honey from southern Spain carries the fresh grass tang of fine olive oil, as if the flowers had found a way to transport their very souls into the nectar.

In the mountains of Croatia, we stopped and bought a jar of honey that tasted like smoke; in the ruins of Jerash, we smeared honey over bread baked in an oven that had done nothing but bake this same type of bread for five hundred years. Same stuff, at least in name, but very, very different experiences.

Maybe the way to phrase it is like this: honey is memory, the landscape's own memory, as measured as a tree ring, as detailed as the pinfeathers on a just-fledged bird. In a jar of thick, nearly black honey, I taste a night in Morocco when I was led through the darkening souk, iron gates slamming closed behind us, the day's final prayer nothing but an echo by the time a gnarled hand reached out through a doorway chink and traded me an unfiltered jar of desert honey for a small handful of soiled dirham notes. And in that honey hides not only the flavor of that night, but also the scent of the sea coming miles over the sand, the rumble of donkey carts and cheap motorcycles, the call of the snake charmers, the tang of wood fires that burned in mountain crevasses, illuminating curls of sleeping plants awaiting morning's first flight of bees.

All in a single drop.

Once you begin noticing these different tastes, colors, scents, and textures of honey, the landscape becomes more and more alive, personalized in a way previously inaccessible, like a whispered secret. Each flash of color, flower or bud, becomes potential

The color of honey varies from pale gold to rich amber,
depending on the species of the plant.

flavor, a miniature drama; and trees that had once merely seemed like nice spots of shade suddenly seem to have music all their own, buzzing as if they themselves were hitting this harmonic, the clear, high note of bees hard at work, offering the tribute of collection and attention, a lover at the feet of the beloved, the yellow body offering a grace note to the flower's own spectrum.

Although in our over-sugared age, honey has been demoted to something smeared on toast, squeezed into a glass of ice tea, flavor as unnoticed as change that's slipped between couch cushions, time was, honey was sacred. The ancients considered honey a kind of magic: treasured for its purity, the fact that it never spoils, never corrupts. The thick fluid was not only the sweetest flavor they had found in nature, but the taste closest to the perfection of the gods—gods who drank nectar in some cultures, the same substance the bees went after.

Honey is the culmination of what the entire earth—rock, soil, water, plant, animal—can do when everything works together, catching a delicate scent that floats on a summer breeze and turning it into clean flavor.

That long-ago day when my father brought home our hives, our bees were destined to be mostly sage and alfalfa bees, bees who worked the neighborhood gardens and caught the fragile blooms that came up in the wash every year after flood season. Flavors coming from no farther away than I was likely to ride my bike on a summer afternoon.

But what my father brought home that night was a bigger gift than he could have imagined: he brought us a discovery of our very home. He brought us the truest sweetness of the world.

By the time I fixed my cereal on the morning my family became beekeepers, the hive was already in business, turning the landscape of our neighborhood into food.

And I was destined to spend the rest of my life chasing honey.

TWO:
The Magic

DEPENDING ON WHOSE count you believe, the world holds somewhere between 16,000 and 20,000 species of bee. Maybe seven make honey. The stuff is that impossibly rare.

Oh, a few other insects make honey as well, or at least something that gets called by the same name—a couple kinds of wasp, who keep theirs in hives much as bees do, and some ants, mostly in the subfamily Formicinae, who actually stash honey in the abdomen of specially raised members of the colony. These living storage units hang upside down from the roof of the nest, regurgitating honey on demand to hungry colony members.

But it is with the honey bee that the true miracle occurs.

Of the bees who make honey, one might name *Apis cerana*, *A. dorsata*, *A. floria*. All have their place in the sweet relationship. But what we usually think of as the honey bee is *Apis mellifera*. This species gets called the western honey bee, but its origins have nothing to do with the West. Despite being now common across the continent, *A. mellifera* is not native to North America; the species evolved on the warm shores around the Mediterranean and was brought to this continent by the Spaniards sometime during the seventeenth century, men who called honey "liquid gold."

In the earliest record we have of honey, a painting in Araña Cave, in the Valencia region of Spain, a man climbs a high cliff on

Bee on a rose at California's Huntington Library and Botanical Gardens.
By the Middle Ages, honey was known as "the soul of flowers."

a rickety ladder. Slung around him is a basket to hold the combs he's pulling from a hive while bees swarm around his head.

Magic always comes at the price of danger.

And surely honey had to be magic. Without science—perhaps even with science—how can one possibly understand the transformation that changes the essence of a flower into perfect sweetness? Our ancestors knew that larger forces than simply a tiny insect had to be at work. L. L. Langstroth, a man who singlehandedly changed the art of beekeeping in the mid-1850s, wrote in *The Hive and the Honey-Bee* (1853): "The Creator may be seen in all the works of his hands; but in few more directly than in the wise economy of the Honey-Bee."

Ancient Egyptians believed bees grew from the tears of Ra, ruler of the gods. Bee authority Hilda Ransome refers to this conviction in *The Sacred Bee in Ancient Times and Folklore* (1937), recalling the familiar declaration: "When Ra weeps the water that flows from his eyes upon the ground turns into working bees. They work in flowers and trees of every kind and wax and honey come into being." As early as 3500 BC, the symbol of the king of Lower Egypt was a hieroglyph of a bee, a symbol that lasted four millennia or more, and meant "he who belongs to the bee," as well as "the beekeeper." Already in the first dynasty, the court held an official position for the "Sealer of the Honey." A relief in the Temple of the Sun, dating from around 2600 BC, shows a man kneeling before a hive of bees, taking the harvest.

Honey was so valued that it was a standard demand of tribute from the lands Egypt conquered, and the Pharaoh in turn rained this honey down upon the priests, who used it in religious rites as a symbol of purity, and who fed honey-sweetened cakes to sacred animals. Many early Egyptians believed that after death, the soul

An antique skep from Belgium and an old bee smoker. Until L. L. Langstroth came along with the removable frame hive, not much had changed in beekeeping for hundreds of years.

took on the body of a bee. Wax from the combs was used to make figurines that could serve as either blessing or curse. According to Ransome, honey even served as part of marriage vows: "I take thee to wife," she records from an ancient tablet, "and promise to deliver to thee yearly twelve jars of honey." When Howard Carter, author of *The Tomb of Tut-ankh-amen* (1927), first went into the boy king's crypt, he noted that beeswax was used as an adhesive to seal the covers of alabaster vases found in the tomb. Tut's after-life must have smelled quite nice. Another famed Egyptologist, E. A. Wallis Budge, reported in *The Mummy* (1893) that bodies were frequently preserved in honey, and he told the story of a tomb robber, thinking he had found a jar of the stuff, only to also find hair coming up on his fingers as he dipped them into the honey for a taste.

In India, Vishnu, described in Swami Chinmayananda's 1980 translation of the *Vishnu Sahasranama* as "the All-Pervading essence of all beings, the master of and beyond the past, present and future, the creator and destroyer of all existences, one who supports, sustains and governs the Universe and originates and develops all elements within," was also described as "honey born." The suras of the *Vishnu Sahasranama* say that "the Lord gives joy, just like honey, one who is not knowable by the senses, one who can cause illusion even over other great illusionists, one who is ever busy in the work of creation, sustentation and dissolution." In Greece, bees led worthy pilgrims to the oracle at Delphi, a seer who likely drank mead laced with hallucinogenic plants to help bring on visions. Although evidence shows the Greeks were a little late to come around to the idea of beekeeping—of course, they had no need, the land was perfect for wild bees and the honey was there for the taking—when they finally got into it, they really got into it: by around 600 BC, laws were suggesting people try to keep their beehives at least 300 feet from each other to prevent the *Apis* version of a traffic jam.

The Greeks, of course, laid the groundwork for the West's mythology of bees, as Ransome describes in some length in *The Sacred Bee*. Zeus was fed on honey from sacred bees as an infant, when his mother hid him away from danger on Crete. While he was there, some honey hunters came in to take the wild product, but when they saw Zeus, their armor burst, leaving them unprotected, and they were soon swarmed by bees.

Our standard honey bee even gets its name from a Greek story, a version offered by Virgil's *Aristaeus* in *Georgics* (quoted here from John Dryden's 1697 translation). We can more or less call

Bees have been around since the Jurassic period, between dinosaurs and mammals, when the earliest flowers attractive to them began to grow.

Today a common part of day-to-day meals, honey was once revered as food of the gods.

Aristaeus the god of honey, although, despite being son of a god—Apollo—and the nymph Cyrene, with whom Apollo fell in love when he saw her fighting a lion to save her father's sheep—Aristaeus is technically a "culture-hero," not a god. He also ends up in deep trouble for using bees to try to seduce Eurydice, wife of Orpheus (who, famed for his singing, was called "honey voiced"), using honey as bait. To appease the gods, Aristaeus sacrificed four bulls, from which bees flowed.

But that was all to come later in Aristaeus' honey-dominated life. From the beginning, he was fed on nectar by the Hours, the nymph Melissa, and the centaur Cheiron, but the trick here is not so much Aristaeus as the nymph. That tradition goes back to Melisseus, mythological Greek king of the Cretans, the first who sacrificed to the gods. His two daughters, Amaltheia and Melissa, fed the young Zeus on goat's milk and honey. From Melissa came not only the name for bees—mellifera, carriers of sweetness—but also a cult of priestesses "fed on honey fresh, food of the gods divine."

From Babylonia to the Indus River, honey was the food of the gods, whether it was Zeus eating nectar and drinking ambrosia, or the deities of the *Rig Veda* with their goblets of *soma*. And all those Viking bashes, toasts to Valhalla, where would they be without the glasses of mead? First thing that happened after you died and reached the great drinking hall of the afterlife was that the Valkyries would hand you a big cup of mead.

If the gods liked honey, then surely, people thought, giving it to them was a good way out of a jam, as serious as a prayer. Honey was the first thing Sophocles' *Oedipus Rex* (429 BC) tried to use to bail himself out of his predicament, told by the Chorus:

An expiation instant thou must make
To the offended powers whose sacred seat
Thou has profaned.
　　　　　　　But how must it be done?
Take thou a cup wrought by skillful hands,
Bind it with wreaths around . . . Then turning to the sun
Make thy libations . . .
The water from three fountains drawn; and last,
Remember, none be left . . . Water with honey mixed—
No wine—pour this upon the earth.

And if it was good enough for the gods, surely it was the perfect first meal for humans as well: across the Mediterranean to the Indian subcontinent, people dabbed honey on their babies' lips in hopes for a rich and contented life. The tradition continues, in different forms, even today: some Jewish sects make sure to have honey as part of their meal on the first day of the year, so that all the days that follow will be as sweet.

Running parallel to all the legend is the fact. Perhaps it was the Greeks who first began attempting to make some kind of systematic study of bees, although even the first scientific writings on bees were full of wonder and speculation. Aristotle began his study of the creatures and makes a number of observations about bees in his *History of Animals* (343 BC) on the same assumption that appears in Virgil's *Georgics*, an idea that goes clear back to the Aristaeus legends, that bees spring whole from the bodies of slaughtered oxen (a belief that continued, and was experimented upon, far into the Middle Ages). Aristotle was also pretty sure that bees didn't actually make honey, but simply gathered it like dew from the leaves. Honey, he claimed, precipitated from the air when rainbows descended.

Think about that for a second: honey as a by-product of rainbows.

The coming of the Age of Reason didn't change the way bees fascinated and, by the secretive workings inside the dark of the hive, confused. Pretty early on, close observers realized one bee was in charge of the whole shebang, but they figured that bee had to be the king. Centuries went by until microscopes were good enough to magnify the dissection of bees and put that kingly misconception to rest, around the mid-1600s. And even afterward, there remained decades of speculation as to how the queen became impregnated, before someone noticed her mating flight, the one and only time in her life of toil that she left the hive.

It took another hundred years or so for people to figure out that the bees didn't gather wax, but produced it themselves; the rebellious North American colonials were starting to glare across the ocean at England by the time systematic studies of comb were completed. And even today the work goes on, as honey's medicinal and curative properties are only just coming into focus.

So maybe it makes sense we have left most of our descriptions of bees to the magicians. In Shakespeare's *The Tempest* (1610-11), Ariel says,

Where the bee sucks, there suck I;
In a cowslip's bell I lie;
There I couch when owls do cry.
On the bat's back I do fly
After summer merrily.
Merrily, merrily shall I live now
Under the blossom that hangs on the bough.

And when the sprite is at last set free from slavery, the bees serve as a representation of everything good in the world.

But Shakespeare wasn't quite done with bees; to him, they were a symbol of everything right with the world, an analogy for life itself:

For so work the honey-bees,
Creatures that by a rule in nature teach
The act of order to a peopled kingdom.
They have a king and officers of sorts;
Where some, like magistrates, correct at home,
Others, like merchants, venture trade abroad,
Others, like soldiers, armed in their stings,

OVERLEAF: *Bricks of beeswax. In ancient times, beeswax—used for everything from candles to writing material—was more prized than honey itself.*

Make boot upon the summer's velvet buds;
Which pillage they with merry march bring home
To the tent-royal of their emperor.

The magic crosses every border. In Transylvania, they leave the house keys out where bees can fly over them, showing the bees how welcome they are to make them more industrious. Prayers for honey were common and sincere. A Bulgarian folk story has bees and their conversations with God responsible for the creation of mountains and valleys, a way of making the sky fit better over the earth. Russians tell a story of Christ changing bread to honey, surely a more useful miracle than the standard water into wine. In China, beehives were turned around after the death of the bee-keeper, as if to help them adapt to the changes in the world.

Virgil's *Georgics* includes this beautiful passage:

Such is their toil and such their busy pains
As exercise the bees in flowery plains
When winter past and summer scarce begun,
Invites them forth to labour in the sun.
Some lead their youth abroad, while some condense
Their liquid store, and some in cells dispense:
Some at the gate stand ready to receive
the golden burden, and their friends relieve;
All, with united force, combine to drive
The lazy drones from the laborious hives,
With envy stung they view each other's deeds;
The fragrant work with diligence proceeds.

Let's leave with this, though, a tale from the dark forests of Germany, where the Brothers Grimm, done with Cinderella and Little Red Riding Hood, turned their attention to "The Queen Bee" in *The Complete Grimm's Fairy Tales*. Two of the king's sons left home to find fame and fortune, but quickly fell in with a bad crowd. The third son, a dwarf, went to straighten them out. The older brothers just laughed at him, but finally agreed to travel with him. Along the way, they came to a castle, where everyone had been turned to stone. A wizened old man showed them the way to end the enchantment, a sequence of tasks. The third task is described as the hardest. It was to choose the youngest and the best of the king's three daughters. Now all the daughters were beautiful, and they were all exactly alike, but the dwarf, Witling, was told that the eldest had eaten a piece of sugar, the next some sweet syrup, and the youngest a spoonful of honey; he had to guess the daughter who had eaten the honey. Then came the Queen of Bees, who he'd earlier saved from the fire, and she tried the lips of all three. At last she sat upon the lips of the one that had eaten the honey, so Witling knew which daughter was the youngest. The spell was broken, everyone returned to their proper form, and Witling married the princess.

Clearly, the bees recognize those who treasure them.

And for those who understand the treasure of honey, the rewards are endless, the magic a miracle upon request, there for the asking, with the simple opening of a jar of honey.

THREE:
In the Hive

NO WONDER HONEY itself was thought miraculous, because each honey bee is a gram of utter miracle. The 60,000 or so bees that make up the average colony have communication systems, air conditioning, and food storage. They may gather nectar from miles away, come back, do a little dance near the hive's entrance, and instantly the other worker bees know exactly where to go. Bees live in the most organized of societies, each knowing the task to perform, and of course, the phrase "busy as a bee" has entered our language for good reason. You can observe nearly any other creature in times of daylight repose; bees, though, as long as the sun is shining, are all about the work, and it is simply our good fortune that their work results in the pollination of flowers and in the delicate substance of honey.

Bees have been on Earth for more than 100 million years, showing up during the late Jurassic handover between dinosaurs and mammals. Why then? For the simple reason that around that time, the earliest flowers that might interest bees also began to evolve. Of course plants had been around—they'd been the dominant life form for 300 million years before the first ancestor of a rose opened on a Cretaceous morning. But then nature decided to change things around a bit, to protect the ovule of the plant with a splash of color, to increase the chance for survival of a particular plant line by adding the diversity of sexual reproduction.

A drone—a male bee—emerges from the cell. Drones make up a small percentage of the bee colony, and serve only to fertilize the queen.

Hence, the flower. And flowers needed something to move that pollen around.

Which brings us to the bee. Not the only pollinating animal, of course: birds, moths, butterflies, beetles, wasps, dozens of other insect species pollinate. In the deserts of the American Southwest, the Saguaro cactus (*Carnegiea gigantea*), that quintessential plant that looks like nothing so much as someone being robbed by bandits, arms held high, is pollinated by bats.

But bees are the only one of these pollinators who have figured out the art of the massive by-product, the creation of huge surpluses of honey. A reason for our closest attention. Meanwhile, flowers have evolved to find ever more intricate ways to draw the bees in to do the job of pollination: from the shapes of the petals to the curve of the anther, even the fact that in some species, pollen is released in response to the vibrations a bee makes while buzzing. Perfect symbiosis.

The first bees were solitary animals—and solitary species still exist, such as mason bees, which hollow out nests for themselves in everything from dead plants to hay-bale-constructed houses. In fact, most bee species are still loners, hardly noticed by humans at all.

But from that initial line of solitary bees, about 80 million years ago, some decided to hang out together, and the splinter groups flowed, including stingless bees and bumblebees. Recent genetic studies have shown that the evolution of the honey bee was considerably slower than that of some other common insects, such as fruit flies and mosquitoes; bees also evolved considerably more complicated behavior, and the genetic indicators show that honey bee genes are more similar to those of vertebrates than those of other insects. This helps explain the fact that bees have circadian rhythms, among other scientific marvels. Bees can also learn easily, and show clear displays of memory. Yet, surprisingly for an animal that gives us the taste of the world, bees have fairly simple taste receptors themselves.

All that evolution leads to a singular event: around 65 million years ago, with flowers blooming, the honey bee appeared. Ever since then, the two—bee and flower—have danced together.

The *Apis* line, those bees who were going to turn their world into flavor, first began to appear with *A. dorsata* and *A. florae*. According to Eva Crane, doubtless the greatest bee expert of modern times and author of *The World History of Beekeeping and Honey Hunting* (1999), *A. dorsata* and *A. florae* probably began building single-comb nests in the open.

Apis mellifera, our honey bee, was yet another branch off the line, evolving in the warm climates of Europe, building its combs and nests in cracks and crevices, learning how to survive winter by huddling together.

Apis mellifera is one of the great success stories of the insect kingdom. From its original home ground of the Mediterranean, it has—with a little help from beekeepers—spread out to nearly every corner of the world, making honey from Greenland to the tip of South America.

And along the way, honey bees have shared their lives with us. Never, it must be emphasized, have they become domesticated; the honey bee today is as wild as it ever was. *Apis mellifera* has simply decided to live with us.

That decision on the part of the bees to hang out with us did bring up some questions. In the early days of beekeeping, people tried to figure out exactly how long bees lived, how long a colony could survive, how, indeed, bees function together at all. Honey

Looking into the beehive.

is, of course, enduring, but what about those making the honey?
L. L. Langstroth wrote:

> For want of proper knowledge with regard to the age of
> bees, huge "bee palaces," and large closets in garrets or at-
> tics, have been constructed, and their proprietors have vainly
> imagined that the bees would fill them, however roomy; for
> they can see no reason why a colony should not continue to
> increase indefinitely, until at length it numbers its inhabitants
> by millions or billions! As the bees can never at one time
> equal, still less exceed the number which the queen is capable
> of producing in one season, these spacious dwellings have
> always an abundance of "spare rooms." It seems strange that
> men can be thus deceived, when often in their own Apiary,
> they have healthy stocks which have not swarmed for a year
> or more, and which yet in the spring are not a whit more

populous than those which have regularly parted with vigorous swarms.

So what really happens in the darkness of the hive?

The worker bees are all female, and they make up about 99 percent of the colony. The males, the drones, exist only to fertilize the hive's lone queen, and when that's done, usually in a single flight before winter comes on, they're kicked out of the hive.

In all the years my family had bees, I was stung exactly once: barefoot in the backyard, I stepped on a bee, probably one nearing the end of her life, when most bees leave the hive a final time, so as not to burden her sisters with having to dispose of the body.

But if you've ever been stung by a bee on the wing, that's a healthy female worker; the stingless drones rarely come into daylight. The workers are the ones you see making the brush come alive in spring, bringing trees and flowers and bushes into life so that if you catch them at just the right moment, they let off a sound like an old refrigerator, there are so many bees at the flowers.

Female bees spend the first half of their lives, a long three weeks, working inside the hive, tending the queen, creating the wax cells for the honey to come, doing basic maintenance. Then they leave the hive, and begin to dance with the flowers. A worker bee may live only a week outside, or perhaps as long as three weeks, before her wings are nearly worn out from the friction of flying so far and she dies of exhaustion. In quieter, winter months, though, she may live, staying inside the hive as long as three-and-a-half months.

OVERLEAF: *An average colony will have around 50,000 to 60,000 bees, each with their own tasks.*

The drones have things rather easier, at least for the short moment of their lives. They eat, they wait to mate with the queen, and that's about it. Larger than workers, smaller than the queen, stingless, unable to perform the transmogrification of bare substance into honey, drones serve only one function, which is to make the continuance of the colony possible by fertilizing the queen, a function the queen only requires once—although on that one occasion she may mate with several drones— meaning the life of a drone is a prime example of nature throwing a handful of solutions at a very tiny problem. And since the drones are not needed except during the queen's single mating flight, they are simply a burden to the colony the rest of their lives. So exactly how long they live has nothing to do with natural lifespan, but is determined solely by the needs of the queen; it may be as long as a couple months, or they may be kicked out not long after their hatching.

As for the queen herself, over the two years of her lifespan she may lay three-quarters of a million eggs, up to 2000 a day. Really, she lives no life but to ensure the continuance of the hive, shadowed and cared for at every step by an army of her daughters who feed and clean her.

When people first began to look at beehives, it never occurred to them that the big bee was the queen. Somehow they missed the fact that all she was doing, her entire life, was moving from cell to cell laying eggs, little more than a one-bee production line. It was clear that this large bee was in command of the hive, so surely, anyone with any understanding of the ways of the world would declare that this was the king bee.

This conviction began at least as far back as Aristotle in *History of Animals*:

Queen cells in a healthy hive. Inside these cells, which are bigger than those for worker or drone bees, the queens-to-be are fed a diet rich in royal jelly.

The king bees never leave the hives, either for food or any other purpose, except with the whole swarm; and they say that, if a swarm wanders to a distance, they will retrace their steps and return until they find the king by his peculiar scent. They say also that, when the king is unable to fly, he is carried by the swarm; and if he perishes, the whole swarm dies with him.

The idea of the king bee lasted a very, very long time, appearing in books in England as late as the seventeenth century; it was also a trope writers liked to come back to when they needed to suck up to a real king. After all, how flattering to be compared to the ruler of nature's most perfect society? What king isn't going to dish out a few gold coins for that?

But the whole idea of a king bee leaves a few problems. Where do the other bees come from, for instance?

Aristotle thought he knew:

> When it is damp, their progeny multiplies; for which reason, the olives and the swarms of bees multiply at the same time. They begin by making comb, in which they place the progeny, which is deposited with their mouths, as those say who affirm that they collect it from external sources. Afterwards they gather the honey which is to be their food, during the summer and the autumn; that which is gathered in the autumn is the best . . .
>
> After the progeny is deposited in the cells, they incubate like birds. In the wax cells the little worm is placed at the side; afterwards it rises of itself to be fed. The progeny both of the bees and drones from which the little worms are produced, is white.

But at the same time, some students of bees were thinking this may not be quite what's really happening. As early as St. Ambrose in the 300s (600 years or so after Aristotle, but still 1300 years before the idea of the king bee was laid to rest once and for all), writers were mentioning the "mother bee." And, despite the name, she isn't really in charge. Although she's the lynchpin of the entire hive, she's not running anything; she's too busy laying eggs. The queen bee is a biological machine, and little more.

Queens begin as the same kind of egg as a worker bee, a standard, fertilized egg. But in the cell, she's fed a specialized diet of royal jelly, and that triggers the morphological changes that make the rest of her life possible.

Queen bees marked to make them easier to find.

A colony will likely raise several queens at a time, when the old queen seems to be faltering. In general, first one out wins, after killing the others as they remain unhatched in their cells. Occasionally, there may be an actual fight in the hive, but bees are an orderly society, and everybody seems to function better when this doesn't happen.

As soon as the new queen is able, she leaves the hive for the one and only time in her life, on what's known as the "nuptial flight," first documented and understood as late as 1788.

The queen heads out, flying circles around the hive, ten or fifteen feet up. She returns to the hive briefly, and takes flight once again. This time around, the drones know what's going on and follow her, looking for their brief moment of swimming in the gene pool.

Over the course of the nuptial flight, the queen will mate with quite a number of drones, as many who manage to catch up and arrange things before she heads back to the hive. As is so common in the insect kingdom, it's a pyrrhic victory for the drones: the very process of mating is fatal. They die with their abdomens torn open when the end of the mating process leaves their endophallus attached to the queen as the male bee falls away.

Much of this is now academic. Commercial hives rarely allow the bees to raise their own queens anymore. Instead, queens are raised elsewhere—Australia and Hawaii, mostly—and imported to the hive. Australian queens were early linked to colony collapse disorder (a condition threatening bees, identified in 2006), although the final verdict is not yet in.

Rearing queens commercially goes way back. Isaac Hopkins wrote in *The Hopkins Method of Queen Rearing* (1886) that when raising queens, "the main considerations are to develop the queens

Queen cells in Hawaii. Most of the queen bees used in hives throughout the United States come from Hawaii.

in strong colonies, and let the nurse bees have as little to do as possible, that their whole attention may be devoted to rearing the queens." The idea is no different than selective breeding of any other stock animal, and it's now so common that today, the 2008 online catalog for University of California, Davis, offers classes on how to commercially raise queen bees, emphasizing steps including the handling of queen cells, drone production, and the establishment of mating areas.

So the odds are, when you see a colony, its queen came through the mail, shipped in a tiny box marked "livestock." Once, when my father was getting a new queen for the hive, they opened the post office for him on a Sunday just because they wanted that box out of there.

However the queen comes to her position in the hive, her labors are just beginning. She'll spend the rest of her days moving from cell to cell in the comb, laying thousands upon thousands of eggs: fertilized for a worker bee, unfertilized for a drone, the decision of proportion left to some unknown mechanism in the way the bees make the combs for eggs, the drone cells a little bigger.

The queen's every need is met by attendant worker bees. They prepare the cells ahead of her, close them off when the time comes, keep the queen fed, dry, comfortable, and clean. And she gets this kind of treatment because the health of the queen is paramount to the hive; should she start to falter—a signal the workers can detect in the pheromones she releases—the workers begin the process of creating a new queen once more, segregating off a couple of cells and beginning to feed the larvae with royal jelly.

As for the worker bees, or at least the eggs that will become the worker bees, life is a little different. Before becoming a fully adult bee, a few stages have to be gotten through first.

For the first three days, the egg waits in an uncapped cell. Drones and workers are laid in cells that are a little different in size, and the queen can tell before she gets down to business. She drops the egg straight into the cell, so it's vertical, and then moves onto the next.

Over the course of the next two days, the egg will start to move; it'll be at roughly forty-five degrees to the cell on the second day, and horizontal to the bottom of the cell on the third day, when it actually hatches.

Now it's a larva. Larvae don't have a whole lot to do except grow into bees, a biologically exhausting process. Over the next four days, worker bees feed the larva, which goes through a moult every twenty-four hours, each bringing it closer and closer to bee-hood.

By eight days after the egg is laid, the larva fills the cell completely; that's when the worker bees seal the cell off, as its cocoon is spun.

Larvae. At this stage of a bee's life, about all it has to do is grow into a bee; workers feed the larvae at a frantic pace, while the larvae moult every 24 hours.

Even then, the moulting is not done. A fifth moult occurs about three days after the cell is sealed; now the larva becomes a pupa. It slowly changes color, from white to proper honey bee yellow.

And on the twenty-first day after the egg was laid, the new bee emerges, having moulted one final time, into adult form, right before breaking out of the cell.

The new bee goes straight to work, taking care of housekeeping, cleaning cells, and helping incubate new broods for the first three days of her life. The next three, she feeds the larvae the mixture of honey and pollen that is all the nutrition they need to finish becoming bees. Younger larvae take up her next few days, as she feeds them what's called "brood food," rich in protein and produced by adult bees from special glands near their mouths.

Then it is a few days in the honey factory, cleaning, drying, and helping maintain temperature and humidity.

Finally, around nineteen days after emergence, she takes her first flights, learning where the hive sits in the landscape.

And as the third week of her life begins, she heads out to turn the world into the sweetness of honey.

Meanwhile, back in the hive, the entire process repeats and continues. Year after year, through the seasons. The life of the colony is endlessly ordered, each bee adding its contribution to the destiny of the whole. And that even means that sometimes, the life of the colony has to be disrupted because things become too crowded. Then it's time to swarm.

A second queen is born into the hive, and, like a general leading a peaceful rebellion, she takes thousands of her followers on a search for a better place to live.

Swarming is how bees spread, how they were able to take over North America so quickly after their first import, how "killer bees" have come north from Brazil. Swarming is the perfect means of both colonization and population control.

Watching a swarm is like watching the sky come alive. The bees pour out of the hive, fly in a ball that increases in size beyond what one would think a mass of insects could look like. A swarm has a special-effects quality that cannot be denied: the air is simply replaced by bees.

The swarm is almost harmless at this point; the bees are very busy and quite distracted, and have no interest in anything but finding a new home quickly. And this is when the bees are at their most vulnerable, out in the open, the entire colony looking for a place to land.

After a while, the queen finds a spot she likes, and she alights. The other bees follow quickly. A landed swarm may gather on a single branch of a tree, or hide in the crevasse of a cliff. The queen knows her own mind, and has her own criteria.

The classic way of capturing a swarm is to simply lead the bees to the new hive. Set up a new, clean hive near where the swarm has landed; then spread a sheet or something similar between the landing spot and the new home. Shake the tree, knock some bees onto the sheet. Curious, they'll explore, and will quickly find the new hive.

Once it's been approved, the bees simply march into their home. It's a steady procession, not on the wing, but on their six legs, walking sedately but with purpose.

The first weeks for the new hive are a precarious time. They quickly must get to the business of finding food for the hive, keeping the queen safe and productive, laying the eggs that will turn into new generations. The slightest upset can doom the entire hive to starvation or premature collapse.

But soon, the combs are filling with honey and the next generations of bees. The work of pollination goes on. And the bees continue doing what bees have always done, building the landscape through their flights.

The colony of bees, the ancients believed, was immortal, and in a sense it is. Bees have figured out how to keep their home and lives running without any disruption, moving on to taking care of the new queen as smoothly as they did the old, while the foraging worker bees happily continue flying out into the world to plumb the souls of flowers.

The work goes endlessly on.

FOUR:
Dancing With Bees

IF I TYPE THE WORDS "orange blossom," will you see the same picture in your head that I do in mine? Of course not; my image of orange blossoms comes from years of living in the desert Southwest, on property that had once been part of a grove; my orange blossoms come from watching the trees in my yard come into full bloom, until they looked like bouquets, putting out that winter scent that always reminds me of the perfume of the first woman I ever truly loved. And your orange blossoms can't do that, just as mine can't do whatever it is yours do.

But let's take the question of representation a bit deeper, to the question of space and position. I say, "The tree is over there," or, "The tree is three hundred yards ahead, turn left when you see the big boulder," or even "Walk into the sun until the land starts to slant downhill a little, and then look left." What are the odds you'll be able to accurately follow my directions without questions and clarifications?

Yet bees do exactly that on a daily basis. Instead of using words, though, they use a dance. And inscribed in that dance—a matter of waggles and circles and lifts—is geographical information that conveys place as precisely as if the other bees were reading it off a Global Positioning System (GPS), botanical information as accurate as any farmer tasting the first buds of spring and reading in them the fortunes of the entire crop to come.

The queen's long body distinguishes her from the rest; here the queen communicates with another bee via her antennae.

With their dance, bees tell each other where and what good sources of nectar are to such a level of detail that two adjacent colonies can be feeding on two adjacent fields, miles away, none of the bees going to the wrong kind of flower.

Exactly how the bees accomplished this delicate communication remained a mystery for thousands of years, stumping everyone from the first cavemen dipping hands into a plundered hive, to generations of scientific observers.

Aristotle wrote in *History of Animals*: "On each trip the bee does not fly from a flower of one kind to a flower of another, but flies from one violet, say, to another violet, and never meddles with another flower until it has gotten back to the hive; on reaching the hive, they throw off their load, and each bee on her return is followed by three or four companions."

Over the centuries, close watchers would theorize about how the bees were passing word among each other, but it took the research of Karl von Frisch to truly crack the code of the dance, which he described in *The Dance Language and Orientation of Bees* (1967). Beginning in the 1920s, with a remarkably complex series of experiments that involved everything from painting a code of colors representing numbers onto individual bees to setting up a photoelectric recorder to follow the movements of the bees when he couldn't be around himself, von Frisch, with ultimate patience, learned to interpret the language of the bees.

Von Frisch set up an artificial feeding source, a spot reasonably near the colony that he could keep an eye on. Marked bees were shown the place, a dish full of sugar water, and within a half hour, they were back at the hive. If the bees had found the feeding station empty, not much happened; if it had been full, though, if they flew back happy and fed, the bee would let a drop of honey water

Bees around the hive's entrance. A single bee may make a dozen or more trips a day to forage.

appear at her mouth; and others would come and take this from her, as the honey-to-be was quickly distributed into the hive.

And that's when the dance begins.

The first words in the language of bees are always about the potential for sweetness. The bee who found the food source begins to run in a tiny circle, not much larger in diameter than a single cell of the honeycomb. The circle widens a bit, and then she changes directions. "Between the two reversals," writes von Frisch, "there are often one or two complete circles, but frequently only three-quarters or half a circle. The dance may come to an end after one or two reversals, but twenty and more reversals may succeed one another; correspondingly, at times the dance lasts scarcely a second and at others often goes on for minutes."

Bees never dance alone, nor even in areas of the hive when there aren't many other bees around to watch. They pick the most densely populated part of the hive, and their sisters slowly begin to join in the dance, five or six other bees imitating the leader's

Bees sitting outside the hive, working to maintain a constant temperature inside. On hot days, worker bees will gather near the hive entrance, flapping their wings as air conditioning.

movements like a giant snowball dance in junior high. And, the "round dance" completed, the information passed, the bees head out to forage.

Clearly, the dance is a recruitment effort, nothing much more complicated than "I've found food, come grab some." That much had been suspected for years, if not centuries. The livelier the dance, the better the pickings, and the more bees would join in the hunt; and as they returned, as the flowers and food were fully gleaned, the dances became less and less vigorous, until no new bees bothered to fly out.

The big question, though: what exactly was the language encoded in this dance? How did the bees, with unerring accuracy, know *where* to fly after simply watching one of their hive-mates turn a few circles?

Von Frisch suggests the performance is, in a way, multimedia. A huge part of the communication lies in scent; the bee returning from the food source simply smells, to the other bees, like a particular kind of food, like traces that linger when a pretty woman walks by wearing a delicate perfume. The other bees are picking up scent traces of the food source (von Frisch's experiments showed traces of flower scent lingered on bees for as long as an hour after leaving the source), so they know what kind of flowers to zoom in on.

But more information is conveyed. The round dance itself gives a rough indication of distance and direction; and the intensity of the dance says how many sister bees need to head out and begin to gather the nectar.

However, von Frisch quickly proved that color information was not transmitted in the dance; the bees went to specific blue plates in his experiments, not just any blue plate. Similarly, he found that there did not seem to be any information passed on about the shape of the food source.

And so, after a round dance, new bees fly out in all directions, doing a quick search, chasing the scent trail until they lock in on the source. Okay, that makes perfect sense; it's like watching a hunting dog in action. But the round dance is only used for nearby food sources, closer than a hundred yards or so away.

How, then, can bees possibly find food from miles away, sources too far for the random search patterns and delicate scent trails to work? Therein lies the true magic of the bee's dance: the tail-wagging dance, a language as abstract as a painting by Mark Rothko, as exact as a science textbook.

To tail-wag, the bee runs straight ahead across a few cells of the comb; she then turns a semicircle to return to the point where

OVERLEAF: *Also known as Bee Bread, the borage herb is much loved by bees.*

she started, before running forward again and closing the other half of the circle. During the run, she wags her body, tail brushing back and forth frantically as much as fifteen times per second. The farther the run, the farther from the hive the food source lies, the more slowly her abdomen wags.

As with the round dance, encoded in the tail-wag are scent clues as to what the food source is, and how much food is out there. That's the easy part. But how does she convey directions with such precision that, unlike with the round dance, where following bees have to do a quick search pattern, after the tail-wag dance, they fly in a straight line directly to the flowers indicated?

First, the tail-wag dance incorporates yet another multimedia approach, adding sound into the mix. The dancing bee gives off a buzzing sound that a person listening in would be able to pick up. But whereas what we would hear would be a continual sound, the dance song is actually a series of pulses, each about fifteen milliseconds long. The more intense the sound, the more bees come to join in the dance, the more bees will head out to the flowers.

Now the dancing bee has to tell her sisters where to go. According to von Frisch, if she runs across two or three cells, the food is within 500 meters of the hive; four cells indicate around 3500 meters, and four or five cells could mean the food is as far as 4500 meters—more than two and a half miles—away.

But she's not done. The tail-wag dance is not just back and forth, but also around. And that's where the actual directions are encoded.

Von Frisch gives an example:

Bees foraging in almond trees.

When the foraging bee flies from the hive to the feeding-place with the sun at an angle of forty degrees to the left and in front of her, she keeps this angle in her wagging dance and thus indicates the direction of the feeding-place. The bees who follow after the dancer notice their own position with respect to the sun while following the wagging dance; by maintaining the same position on their flight, they obtain the direction of the feeding source.

Intensity, angle, sound, scent. The bees have their own complex language and grammar, all hidden inside a dance not much different than doing the hokey-pokey.

But a final consideration must be made: what time it is. Timing is everything with not only flowers and their production of pollen and nectar, but flight times, the evening sun coming down and the colony's evening dormancy beginning.

A bee gathers nectar from a sunflower.

Von Frisch wrote:

It is now clear that we are dealing here with beings who, seemingly without needing a clock, possess a memory for time, dependent neither on a feeling of hunger nor an appreciation of the sun's position, and which, like our own appreciation of time, seems to defy any further analysis. So far as precision is concerned, it is doubtful if we are a match for the bees, as we can only estimate correctly the length of a fairly short interval, whereas the bee, even in the monotony of the illuminated room, is able to recognize the hour of the day at which it had been fed. . . . The bees' sense of time is of the greatest importance and affects also their orientation in space; for the sun is only of use as a compass if one can tell the time of day.

And their sense of time tells them this: it's time to go outside, into the fields and the flowers, to add their own flash of yellow and black to the colors of the landscape.

Then they'll come back to the hive, where it will be time to dance yet again.

FIVE:
From Nectar to Honey

IN A GOOD, productive area, it takes over 50,000 miles of flying and the bees visiting more than two million flowers to make a single pound of honey. In other places—say, the Middle East, where some of the most delicate honeys come from—the bees may have made the equivalent of a journey to the moon to gather flavor.

The average worker bee, before she drops dead of sheer exhaustion, her wings nearly worn out from the friction of air as she flies up to sixty miles per day, will contribute roughly one-twelfth of a teaspoon of honey to the hive.

Now consider that in the United States alone, we consume more than 400 million pounds of honey a year. It's endless work for the bee, turning color into taste.

So now it's time we do what the bees do, and think about flowers for a while. Take a look at a flower; notice how even if it were the same color as the leaves of its plant, it would stand out, different in shape and form from the background. To bees, flowers are exclamation marks in the landscape, a moment of life running rampant and exuberant.

Or to put it another, more practical way, the flower is nature's equivalent of Pavlov's bell. Those shapes, colors, forms, smells, alert the bee that there is something useful, something desirable in the flower.

Bee deep in the colors of roses.

Yet there's so much more to them than what we see and sense. Bees do not have the same kind of color vision we do—their spectrum is slightly different, and they don't see red. So some of the most popular flowers—roses, say, or poppies—are distinguished by the bees not for their shapes and shades, but rather for their disturbances in the ultraviolet waves, something bees see that we cannot. And what the bee sees lures her like a lover's beckoning finger.

Nectar, at a fundamental level, the thing the bee desires, is a bribe, plain and simple. Plants produce it in order to get animals to come to them, like bait on the end of the hook. The plant needs animals—bees, beetles, flies, even the much-maligned but vital and beautiful bats—to come inside, get a slight dusting of pollen on them, leave behind some pollen picked up elsewhere. Pure sex. Birds do it, bees do it, and plants do it. They just need a little help from the birds and bees.

How this all works comes down to a bit of bee physiology: evolution has put fine hairs on the insect. These gather an electrostatic charge, drawing power from the air itself, and that charge attracts the pollen. The nectar is hidden deep in the flower, ensuring that to get to it, the animals have to brush by the pistil and anthers, the flower's sexual organs. When the bee flies to the next flower, a bit of this pollen gets left behind, fertilizing the new plant in what can only be described as a beautifully contrived accident.

We'll get to the details of this later, but without this simple pollination process, as many as two-thirds of the plant species in the world—and roughly a third of the average U.S. diet—would simply disappear.

A bee already heavy with pollen from foraging soaks up some spilled honey.

The bee, of course, has her own use for pollen. From time to time during the work day, she stops to groom herself, gather the fine grains into the scopa, a structure like a little basket, on her legs. Back in the hive, this pollen feeds the machine that is the colony, grain after grain going to the pupae, who, as they develop, need to be fed every few minutes. And beyond that, pollen is, in effect, what balances the bees' diet from all that sugar of the nectar: it can be nearly 30 percent protein by weight. It's also full of essential amino acids the bees require for health.

L. L. Langstroth, author of *Hive and the Honeybee*, went back to Aristotle on pollen:

> A bee, in gathering pollen, always confines herself to the same kind of flower on which she begins, even when that is not so abundant as some others. Thus if you examine a ball

OVERLEAF: *Bee gathering nectar from an Indian blanket flower.*

of this substance taken from her thigh, it is found to be of one uniform color throughout: the load of one will be yellow, another red, and a third brown; the color varying according to that of the plant from which it was obtained. It is probable that the pollen of different kinds of flowers would not pack so well together. It is certain that if they flew from one species to another, there would be a much greater mixture of different varieties than there now is, for they carry on their bodies the pollen or fertilizing principle, and thus aid most powerfully in the impregnation of plants.

So the bee is doing three things when she lands on a flower: gathering nectar, spreading pollen, and making life on our planet worth living. The honey is just an incredibly luxurious side effect, a gift.

On the flower, the bee eats as much nectar as possible, like a glutton at a banquet, before returning to the hive. In an area thick with useful flora, this may only take fifteen minutes, and, barely pausing to rest, the bee may make a couple dozen trips before her day is finished.

Honey is a chemical process, according to scientists. The nectar that the bees spend their days gathering carries a complex combination of sugars—sucrose, fructose, and glucose making up roughly a third of the nectar's total content. Honey, though, is so concentrated that it is about 80 percent sugars.

Inside the hive, worker bees evaporate the excess water in the nectar, working tirelessly over the blooming honey, which becomes more and more dense, taking up less and less space in the hive, a way of maximizing the caloric footprint. The bees also secrete an enzyme, invertase, which changes the sucrose into fruc-

Tupelo honey, foundation for a Van Morrison song, and arguably the best honey produced in the United States, from tiny blooms in the swamps of the Gulf Coast.

tose or glucose, allowing even more sugar to be held in the liquid. And that alone—the amount of sugar honey holds—is a small miracle, helping keep the honey sterile.

Honey has a few other tricks, as well. In the days before micro-waves, maybe the first magic trick any child saw was watching Mom put a jar of crystallized honey into a shallow pan of water; as the water heated, the honey turned back into pure liquid.

Not all honeys crystallize—another side effect of particular flowers. Tupelo, one of the most treasured honeys, brought forth from the swamps of north Florida and south Georgia, never crys-tallizes. (Tupelo was also the inspiration, of course, for one of the best songs ever written about honey, by Van Morrison.) Fine kiawe honey from Hawaii, made from mesquite and nearly pure white in color, also refuses to crystallize.

But for the most part, crystallization is actually a sign of a more pure honey; the pasteurization process so much commercial honey is subjected to retards crystallization, as if they're afraid consumers will think the honey has spoiled.

Pure honey never spoils. Honey stored for a thousand years will still be perfectly edible. Only adulterated honeys, or honeys thinned with water, can go bad—or under the right conditions, watered honey can turn to mead, something we'll come back to.

A single hive can make 150 pounds of honey over the course of the warm months; in cold climates, they'll use a third of that to get through the winter, but in more temperate regions, they hardly need any honey as a backup plan, as something is blooming all year round and there's always food coming in.

Yet most people never discover the depth of possibility within honey. They go to the grocery store, buy a plastic jar of something labeled "honey," and call it good, even though that jar most likely came from another country, and is likely nothing more than the last squeezings from dozens of hives in dozens of landscapes. The junk food equivalent of honey, if such a thing can be said to exist.

Meanwhile, in the fields, beekeepers go to ridiculous lengths to surround their bees with only single types of flowers, cleaning out every scrap of old crop from the hive before moving the bees to a new landscape. You don't find single-flower honeys on diner tables. They must be sought, discovered like rare gems glinting color in the first bloom of daylight. Linden, sage, heather, eucalyptus, leatherwood—the possibilities are limited only by the landscape itself.

The act of seeking these pure honeys is not without controversy, though. Bees have adapted themselves over countless gen-

Honeycomb, the cells fat with the golden liquid of honey. A productive hive can make more than 150 pounds of honey in a single year.

Clover, the base of the popular honeys.

erations for specific conditions, particular plants, and flavors. The fact that they can come into a new environment and still find something to eat is a measure of their amazing capabilities more than their preferences. We don't know what the bees think. When they're gathering clover in the Southwest, are they feeling some nutritional lack, a gap in the genes that their ancestors had filled with the delicate blossoms of olive trees, or oranges? As bees come under more and more attack from outside forces, from the mites and from the mystery of colony collapse disorder, these questions are being asked: have we pushed the bees too far, taken their gift farther from its source than we were ever meant to?

Make no mistake: the taste of tupelo honey is something as pure and elemental as the chords of a Bach sonata, a perfection of form that simply cannot be improved upon.

But we don't know what the final cost of this genius is, and that leaves us no choice but to respect it and treasure it all the more. To perhaps bow in respect when passing that perfect symbiosis of bee and flower, bringing life and flavor out of color.

SIX:
Living With Bees

"A PERSON WHO KNOWS very little or nothing about bees should make friends with someone who does, and watch him handling the hives and stocks. It is a good thing to find out in this way if one likes bees before starting out on one's own," reads a booklet put out for Britain's Young Farmer's Club in 1943. Back then, in the days before zoning restrictions and endless suburbs, you might have expected to know someone who kept bees.

Since the earliest times, before anyone bothered to write down the history, people have treasured bees as the producers of the sweetest taste they knew. But unlike most of the other animals people have come to depend on for regular food sources—from chickens to cows—bees cannot be domesticated; they can only be encouraged to live near us, and to share their lives and magic.

Honey was a wild food for thousands of years, foraged for on cliffs and in trees. The earliest hominids, four million years or so ago, no doubt learned to follow the buzzing sound toward a food source, and they would have feasted not just on the honey, but also on the pure protein of the larvae.

Being able to find honey was a valuable social skill. Some people could simply intuit where a colony might hide, learning how to think like bees, or, just as importantly, like animals who fed on bees and honey and whose senses might be rather sharper than our own.

Beekeepers work year-round to keep their hives healthy and productive.

Some people could "line" bees—follow a single bee back from flower to hive, no matter how quickly the bee flew nor how indirect the route. Or, since bees tend to follow distinct paths between the hive and the food source, a liner could follow one bee until losing sight, and then simply wait for another to fly by, headed the same way. And so, like an inadvertent relay race, the colony would reveal itself.

And sometimes, of course, people just stumbled across wild hives when they were on other business.

Ownership of wild hives was an important issue; finders claimed their trees, and the claims were easy to uphold in courts of law. Honey poachers were reviled, fined, imprisoned, and, depending upon whose honey they poached, they possibly faced even worse fates, up to and including execution.

Although even today wild hives still no doubt vastly outnumber the "domestic" hives (and, in fact, in plenty of places in the world today, the honey hunt is not much different than it would have been at the dawn of history), let's face it: after a while, people got tired of looking for honey and wanted to bring it closer to home. Which meant bringing the bees themselves closer.

A daunting task, as I remember so clearly from the clean white boxes of our first colony appearing in the backyard.

BEFORE WE GET FURTHER into how humans have found ways to live with bees, let's digress for a moment to think about the architecture of the hive from the bees' perspective. What's required? Room for 60,000 or so sisters; a place for storage, another

A natural beehive—what bees build when left to their own devices.

for child rearing. A clean environment, maintained at close temperature and humidity specifications, no matter what the weather is outside.

In the wild, when bees swarm, the colony sends out scouts to look for a suitable place to live. Like any prospective homeowners, their demands are pretty set. They want something protected, safe. They want a space proper for the size of the family—according to Eva Crane, wild bees look for nooks and crannies that are around twenty to one hundred liters in size. And when they've found a good spot, the entire swarm moves in, looking like nothing so much as concert-goers headed back to the parking lot after a show, a stream of life all moving in one direction.

Inside the hive, the bees set up their furniture: the combs and cells. And they are very, very precise about this, because time and evolution have perfected their understanding of their needs. As Langstroth wrote:

> The cells of the bees are found perfectly to answer all the most refined conditions of a very intricate mathematical problem!
>
> Let it be required to find what shape a given quantity of matter must take, in order to have *the greatest capacity, and the greatest strength*, requiring at the same time, *the least space, and the least labor* in its construction. This problem has been solved by the most refined processes of the higher mathematics, and the result is the hexagonal or six-sided cell of the honey bee, with its three four-sided figures at the base!
>
> The shape of these figures cannot be altered, *ever so little, except for the worse.*

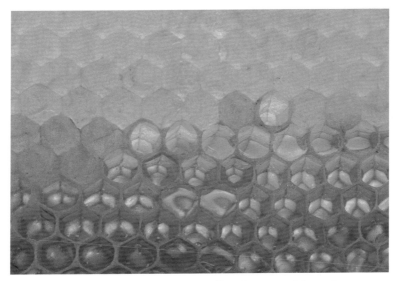

A close-up of honeycomb. Bees came quite naturally to the octagonal shape, the most efficient use of space for storage in the hive.

And what all that means is that people, if they are going to give the bees places to live, must either let the bees design the interior themselves, or they must give the bees something they can simply move into with no hesitation, like a business traveler settling into a hotel room after a long day of meetings.

The first domestic hives were probably nothing more than sections of logs, simply lifted from the wild and brought close to the settlement. The bees had already done the work, and so it was just a matter of moving them where you wanted them.

From there, it wasn't a big leap to simply making things that were close enough to logs to keep the bees happy: ancient Egyptians made hives of mud; on the Arabian peninsula, they made hives of pottery; and in Spain, they rolled cork into the shape of a log and gave the bees a home that way.

All of these hives shared a similar problem: they were heavy, big, and relatively expensive to make, both in time and effort. So

something different was needed, and that, it turned out, was the skep. A skep is a simple twist of weave and straw, curled into a dome. Although designed only to last the bees a season and no more, it has somehow become the emblem for bees, a universal sign immediately recognized.

Making a skep is like making a coil pottery dish in grade school art class, except you go at it from the top down, rather than the bottom up. Gather straw, bind it into a ropelike shape; the thickness is entirely a matter of personal preference and practice. Now coil the rope as tightly as possible; this becomes the top of the skep, a form that widens toward the bottom. The layers are tied together, stitched with string or even more strands of straw, to hold the form steady.

To make a hole for the bees to come and go, simply reverse direction on a couple of the coil rows, before swinging around in a complete circle again. Keep in mind that bees like close spaces; a small, narrow entry helps the bees keep the hive free of predators.

And that's it. Close off the bottom, and the skep is done. You've just made the same form of beehive that has served humans around the world, and is still in use in many places (although illegal in many parts of the United States, due to the same kind of pointless regulations that prevent us from importing cheese with actual flavor).

But the perfection of the form didn't mean that people were done finding ways to live with bees. What does it mean to live with bees on a daily basis, to make bees a part of the family?

An hour or so by train outside Ljubljana, Slovenia, in a tiny

A skep, one of the oldest kinds of man-made beehives: coils of straw rolled into an igloo-like shape. Still common elsewhere in the world, the art of the skep has nearly vanished in the United States.

hilltop town where the main square barely has room for sleeping dogs, is a beehive museum called the Museum of Apiculture. The Slovenes liked to decorate their hives, which are simple boxes, not shaped much differently than a fallen tree trunk. But because the bees were brought near the house, because the bees themselves were seen as something nearly divine and sacred, the boxes were decorated with religious motifs. "The very first examples were undoubtedly of apotropaic character," reads the museum's translated brochure, titled *Pripovedi S Panjev*, "since we know that in other lands as well beekeepers protected sensitive bee families against imprecations and misfortune with special apotropaic signs and even with special blessings. These protective signs could be very rudimentary: a cross, trotamora—a star-shaped symbol warding off evil, Christ's and Virgin's monogram. There was only a short step from this practice to the depiction of the entire saint on the front board of the beehive."

But the brochure's description does not do the hives justice. The Slovenes gave the bees homes behind fine works of art: a painting of the Last Supper, or Eve tempted by the snake, with a unicorn looking on somewhat less than pleased. Souls being weighed, demons and angels at the shoulder of a man trying to make a decision.

Not all the paintings are religious. One shows a man fishing with a pair of pants, three women reaching for them. Surely that makes sense to someone. Snails drip blood from fangs, attacking men trying to climb trees to escape; foxes shave a hunter, illustrating the Slovene equivalent of the proverb "to pull someone's leg."

The museum has a couple other fairly unusual homes for bees as well, showing more of the different ways people have chosen

Preparing frames, the basic structure of a man-made beehive, invented by L. L. Langstroth in the mid-1800s.

to live with bees, including a life-sized carving of a janissary, an infantryman from the Ottoman empire; the body is hollowed out, the bees entering and leaving the hive from, well, a rather unusual aperture.

The hives served a dual function: first, clearly, to protect the bees through divine or profane symbols, to reflect the bees' own near divinity (everyone knows bees all buzz together at exactly midnight on Christmas Day); but also to bring the bees closer to the house, to take them in as part of the family. You hang pretty things on the walls of your home; why should it be any different for the bees?

So much care was taken with the hives because the bees themselves had to be taken in, made members of the household, or else their tacit agreement of sticking around, milking the souls of flowers simply for our delight might be broken.

Bees had to be informed of a death in the family; if they discov-

Beekeepers pull forms from the hive to check on the bees.

ered their original keeper no longer came by, they would stop producing. In England, the bereaved family would tap three times on the hive and tell the bees of the change of keeper; this was so important, the ability to move the bees from one person to another, that it was considered nearly unthinkable for a single person to own hives.

Beekeepers had to acknowledge swarms with an announcement known as "tanging," when the beekeeper was supposed to make a noise with a pan, kettle, mortar, whatever was at hand. The sound was thought to help keep the bees nearby, show them they were welcome. Really just a practical measure, something to tell the bees you knew and understood what was going on.

The swarms had to be welcomed, because if they weren't, they could become quite bad things. In some parts of England, a swarm landing on a dead tree meant there would soon be a death in the family, but in Wales, if the swarm landed on your house or buildings, it was a sign of prosperous days to come. Although the Austrians would agree with the Welsh, they would add that if the swarming bees fight (swarms are usually very, very docile) or try to go back into the hive, it's a sign of impending war.

In Ireland, early laws levied fines for carelessly killing bees, and the mead-brewer was one of the most important men in any town; frequently, he was paid in a substance even more valuable than cash: the beeswax itself. The Welsh considered bees of divine origin, straight from Paradise—and so when the Catholics came in with their beeswax candles for worship, it made perfect sense.

England's own earliest important record, the *Domesday Book*, the first survey of England's lands and people done at the behest of William the Conquerer in the eleventh century, is full of bees and honey. The land was so watched over that even wild honey was appraised for value, and those who could not pay their taxes in money or other goods were more than welcome to pay in honey, which was valued around thirteen pence a pint, a fairly serious chunk of cash.

An imperial symbol of Poland, part of their crown, is a bee, to show the importance of society working as smoothly as a hive, and to commemorate that during a succession battle, the king of the country was chosen when a swarm of bees alighted on one of the contenders. And there were bees in the regalia of Napoleon.

Again, we have to recognize that bees are not, in any sense of the word, domesticated animals. They have merely agreed, for their own reasons, to live near us and share their honey. Bees in clean white-box hives, being moved on the back of flatbed trucks across interstates from field to field, are absolutely no different than wild bees; they are every bit as wild. They merely live in different houses, completely indifferent to anything but the task of their endless work.

And so, beyond the question of living space, how have we chosen to live with them?

The writer Columella, an encyclopedist in Rome during the first century AD, laid out the full work of a year for the beekeeper in *De Re Rustica* (68–42 AD). March is the time for purifying the hives, making sure they're free of moths and spiders. Come April, one hunts for colonies, preparing to catch May's swarms. June is the first honey harvest month; October is the next, but the beekeeper must be careful on the second harvest to leave enough for the bees to make it through the winter. And come November, it's time to prepare the bees for the cold to come, the hives cleaned and protected. "A position must be chosen for the bees facing the sun at midday in winter," he wrote.

The first honey bees came to the Americas in the early seventeenth century. Once the complications—keeping a hive alive during the month-long sea crossing—were solved, bees spread out across the new landscape as if it had been made just for them. Owned hives were only a tiny part of the spread of bees in North America; as colonies grew, the bees swarmed, finding endless food supplies, pushing out their own boundaries.

Bees had spread across Massachusetts by the 1640s, were thriving in the Carolinas by around 1700, and were on the banks of the Mississippi River by 1750. It took another fifty years for them to reach the banks of the Missouri, and they were spreading across Texas by the 1820s. Bees moved just ahead of European settlers, who were in the middle of their own swarm across the landscape. For the Natives, bees—remember, *Apis mellifera* is not a North

Winterizing hives. In cold weather, bees nearly stop their activity, huddling together for warmth, feeding off their honey stores.

American insect at all—were a harbinger of what was to come. They were "the white-man's flies," something that moved as much in tandem with the settlers as the cattle. Longfellow brings this out in *Song of Hiawatha* (1855), which is fairly hackneyed by modern standards, but still has its own ring of truth:

> Wheresoe'er they move, before them
> Swarms the stinging fly, the Ahmo
> Swarms the bee, the honey-maker;
> Wheresoe'er they tread, beneath them
> Springs a flower unknown among us
> Springs the White Man's Foot in blossoms.

Historians generally concede that the first full-time beekeeper in the U.S. was Moses Quimby, of New York State, whose beekeeping thrived around the time of the Civil War, but the odds are, dozens of other beekeepers simply got lost to the records, quietly tending their hives, bottling the flavor of the landscape.

What this all really goes to show is how marvelously adaptable bees are: they find a place, and they find a way to fit in. So maybe now is a good time for a story from the other end of the world, on the Faeroe Islands. How a place that never had bees has come to live with bees.

To slightly rephrase Ringo Starr, to find the Faeroes, you go to Greenland and take a right. The other choice is to head north into Scotland and keep going until the ice stops you. The Faeroe Islands are rough and rocky, offering little but the face of cliffs to the sea. But people have found a way to live here, descendants of the Vikings, fishing and raising sheep and a few vegetables.

Right before World War II, a woman named Hansina Justinussen decided she wanted a fresh apple. She lived in the largest town on the Faeroes, Torshavn, where apples were a rare and expensive imported luxury. But Hansina had an endless fascination with botany and science, so she decided to experiment. She brought a couple apple trees and planted them in her yard. Easy enough. They had to be kept warm, and out of the wind, but the trees grew and bloomed.

The problem was, no bees on the Faeroes, and no way to pollinate.

So Hansina did it by hand. She made a little feather-duster kind of instrument, and brushed pollen from one tree to another. And sure enough, a single apple started to grow. She watched it, all through the long, bright, arctic summer nights.

And the day before she planned to pick it and eat it, someone stole it.

That's life without bees.

But the Faeroes have bees now, proof that, given a chance, bees can thrive anywhere. A few decades ago, the country went through a soccer craze. To get other teams to come play, though, they needed soccer fields, and that required grasses that weren't exactly native to the Faeroes. So the islanders imported turf. And in some of those bundles of turf, bees flew out, found a way to create a home on the island, a place where apples are still scarce, but possible.

In the past twenty or thirty years, beekeeping has become an almost invisible art in the United States. Would-be backyard beekeepers often are prevented from having hives, because of zoning

OVERLEAF: *Honey is the truest taste of the landscape; here, surrounded by bees, a beekeeper looks over the countryside his bees will turn into food.*

Bee hives. Commercial beekeepers may have a thousand or more hives.

restrictions and ignorance on the part of people who fear the mere sight of a bee. Largely owing to bad press and years of stories about killer bees, any wild swarm is likely to be killed the instant it lands within a hundred yards of people.

Still, in some cities, urban beekeeping thrives. Paris, New York, and Chicago all have rooftop apiaries, the bees somehow finding ways—flower boxes, parks, patches of color that a passer-by would never notice—to thrive.

But more, the work of the hives continues in two ways: wild and deep in the forest, the mountains, the deserts, the landscape of the country that has so welcomed them from the beginning; and in the work of the nation's commercial beekeepers, who may truck their hives thousands of miles over the course of a season, fertilizing everything from berries in New England to the Deep South.

The simple truth is we can't live without bees, but it is imperative that we find new ways to make it easier for them to live with us. Without their gifts, without the fine care of beekeepers, the world becomes a much bleaker place.

SEVEN:
The Sting

BEES ARE NOT exactly defenseless. Mention bees, and the first thing that leaps to most people's minds is the pain of their sting. A purely defensive weapon, a sting can be nothing more than a quick, painful annoyance to some people, whereas for others it can be life threatening. Even members of the same family can react very differently: I say "ouch," and get on with life. My brother swells up like a balloon and spends days in bed. Stings can even kill, sending people into severe anaphylactic shock, closing off the throat, causing edema, dropping blood pressure down toward zero.

For the bee, it's no fun, either. Administering a sting is a death sentence for the bee herself. Only the worker bees can sting. The stinger is barbed, like an illustration of a harpoon in an old edition of *Moby Dick*; when the bee flies away, the barbs stay in, tearing out her abdomen as she goes. Unlike, hornets, which can sting repeatedly, a bee can only sting once, and for an average honey bee, it takes a goodly amount of provocation before she'll even think of stinging, especially if she's not anywhere near the hive; up close, they may swarm out to defend, but in the fields, alone, the sting is an absolute last resort. A bee at a distance from its hive never volunteers an attack, the classic writers say.

And if a simple "live and let live" credo weren't enough, folklore provides plenty of ways to avoid being stung at all. In 1770,

*It takes significant provocation to make a bee strike; its first sting is its last,
killing the bee as it flies away.*

A calm beekeeper is only rarely stung, and bees will land on, walk around, and even make eye-to-eye contact with their keepers.

Thomas Wildman wrote an echo of the understanding of bees in the larger world around them, of bees as moral forces, in his *Treatise on the Management of Bees*, which summed up ideas that went as far back as the Roman writer Columella and beyond:

> If thou wilt have the favour of thy bees that they sting thee not, thou must avoid such things as offend them: thou must not be unchaste or uncleanly; for impurity and sluttiness (themselves being most chaste and neat) they utterly abhor; thou must not come among them smelling of sweat, or having a stinking breath, caused either through eating of leeks, onions, garlick, and the like, or by any other means, the noisomeness whereof is corrected by a cup of beer; thou must not be given to surfeiting or drunkenness; thou must not come puffing or blowing unto them, neither hastily stir among them, nor resolutely defend thyself when they seem

to threaten thee; but softly moving thy hand before thy face, gently put them by; and lastly, thou must be no stranger unto them. In a word thou must be chaste, cleanly, sweet, sober, quiet, and familiar; so will they love thee, and know thee from all others.

Drinking beer to cover up bad breath is an idea that went out of fashion long ago, of course, but one wonders how the bees would react to mead breath?

Finally, when describing how to get along with bees, Langstroth echoes instructions from Columella: "Above all, never blow on them; they will try to sting directly, if you do."

The composition of the venom carried by the sting is quite interesting: fully half of the chemicals in it are devoted to lowering the blood pressure of the victim. Lower the blood pressure and you lower an animal's ability to react. Only a bit more than a tenth of the venom is what causes the pain and toxicity of the sting, and what makes my brother explode from a sting is barely a single percentage point of the venom's total composition. Once again, bees prove themselves the alchemists of the animal world.

If you remain calm, bees can land, walk around on you, and go on about their life after sharing yours for just a brief moment of eye-to-eye contact (and bees have marvelous, compound eyes, seeing in a way we can't begin to imagine).

But that all brings us to the movies, because Hollywood prepared us for this long, long ago: swarms of bees, filling the hori-

OVERLEAF: *Bees are swarming on a rose bush. Swarming is how colonies naturally divide and reproduce, but due to public complaints, this swarm will be removed by a beekeeper. Hopefully, education will help people understand that bees and people must coexist for a healthy ecosystem.*

zon line like a storm cloud, their angry buzzing loud enough to block out the screams of people running like they were trying to get out from under Godzilla's trampling foot.

In 1978, Irwin Allen, the man who brought the world classics such as *The Towering Inferno* and *The Poseidon Adventure* (as well as one of the greatest camp TV shows ever, *Lost in Space*), turned his attention to bees with his movie *The Swarm*. Not just any bees, of course. Africanized bees, "killer bees," a controlled biology experiment that had become very, very uncontrolled. For nearly two hours of trudging film, Michael Caine looks desperate to be absolutely anywhere else while he watches millions of bees do in his fellow stars, including Henry Fonda, Olivia de Havilland (feeling very far from Seven Oaks), and even Fred MacMurray in what turned out to be a sad final role for the distinguished career of a very fine actor.

Of course, the Africanized bees kill them all. Plus a nuclear power plant and the Astrodome. And, by actual count as expressed in the film, more than 40,000 people who chose the wrong time to be wandering around outside. These evil insects finally get their comeuppance in a fire that not only sets them, but most of south Texas, ablaze, while Richard Widmark, who lit the match, solemnly intones, "Will history blame me or the bees?"

Irwin Allen's genius was always to get in on the early side of a wave, and the early 1970s were when the first rumors of killer bees headed north were starting to come into the national consciousness.

But the true story of killer bees had begun much, much earlier. Remember, the honey bee is not native to the Americas; it was brought here hundreds of years ago, and thrived as very few animals have ever thrived, but that doesn't mean the new territory

Preparing queens for shipment by inserting them into cages.

didn't bring limitations to an insect that originally evolved around the temperate Mediterranean basin.

And, of course, as our history proves again and again, humans will never leave anything alone to find its own way and accept its own limitations. So, in 1956, scientists, led by an Englishman named Warwick Kerr, decided to try to cross-breed a species of bee that would thrive in a climate they'd done well in but never entirely gotten used to, the South American tropics. In a fit of demented inspiration, they imported twenty-six Tanzanian queen bees, from the subspecies *Apis mellifera scutellata*, to a facility in southeast Brazil. Brazil was footing the bill and offering facilities because their government had a simple agenda: to take their country from number forty-seven in world honey production and put it in the top ten. To do this (and they have, despite themselves, succeeded over the years—most of the honey you see in supermarkets and restaurants comes from Brazil, although lately that's

A beekeeper carefully inserting a queen into the hive. "Killer Bees" made their first appearance when a South American beekeeper did not properly tend to transplanting queens into new colonies.

being supplanted by Chinese honey), they clearly needed to improve upon nature.

The Tanzanian strain of bees was known to be more aggressive—as just discussed, even when attacking, most honey bees won't venture far from the nest, but the African subspecies has no real problem chasing away potential intruders for up to a mile, and the sheer density and ferocity of the attack means that whereas a similar provocation at a normal hive might result in a dozen or so stings, *Apis mellifera scutellata* responds with overwhelming force, the angered bees likely to inflict hundreds of stings. Hence the "killer bee" nickname, although even with these bees, if you get far enough away during an attack, you should be okay.

The trick is just moving faster than the bees.

But that's the downside of *Apis mellifera scutellata*. In the good column, some studies show they're also better at the vital job of pollination than their smaller cousins. Plus, they are stronger and able to produce more honey than the European species that have worked the Americas for centuries. However, they don't hoard honey, the way the European bees do—that hoard being the surplus that feeds us. Instead, they swarm more frequently, and so spread their territory much, much faster; they're also much more inclined, at the first hint of a diminished food supply, to pack up and move elsewhere. These characteristics make the bees resilient to climatic and ecological constraints.

And so the scientists had what seemed like a pretty simple idea: combine the best traits of the European and African bee through careful genetic selection, and make a super bee, one that was strong and adaptable and made tons of honey.

You'd think the scientists would have seen enough horror movies to know what was going to happen.

In 1957, only a year after the bees were brought to South America, a replacement beekeeper accidentally released the Tanzanian queens, allowing them to mate uncontrolled with local drones. This, in effect, removed the experiment from the experiment. No longer did humans have control over how the bees worked out their own sudden burst in evolution; instead, it was all up to the bees themselves.

And instead of combining for the "best" characteristics, as humans would define such a thing, the two species got together and did indeed create a super bee, but one with only the bees' own agenda in mind: strong, fearless, capable of defending themselves from any predator, able to fill in new ecological niches not previously accessible.

In other words, by doing exactly the opposite of what the scientists had hoped for, the bees absolutely could not have been more successful from the point of view of the biological imperative.

And so, with their newfound strength and abilities, these "Africanized" bees started to expand their territory, taking advantage of the fondness of *Apis mellifera scutellata* for frequent swarms. They came out of the jungles, and found the living got easier.

For quite a while, the scare was tinged with hope: surely we'd find a way to stop them before they crossed the Rio Grande and made it into the United States. Or maybe the climate would simply do them in; since they didn't store a lot of honey, a hard winter might simply kill them off.

That isn't quite what happened. The killer bees crossed the U.S. border around 1990, at the south tip of Texas, and by 1993 were well established across the Southwest, provoking ever-increasing scares in the U.S. media with headlines like: "Beware of Bees," "Bees Sting Ten Hikers Multiple Times: Swarm Agitated After Hiker Falls into Hive," "Agriculture Department Sets Traps for 'Killer Bees'," "Immense Bee Swarm Stings Man, Kills Beagle." The stories are pretty much the same, the variations endless.

It is true that they can't withstand cold winters—at least not yet, although we have no idea how adaptable they might be in the long run. But so far, this has kept them mostly west of the Mississippi, with their northernmost permanent incursions taking them past the casinos of Las Vegas, and, appropriately enough, into the deserts hiding Area 51 and the site of all those 1950s giant bug movies. *Apis mellifera scutellata* have been found in Kansas City, and in summer may appear as far north as New England, although those are transient populations unable, so far, to gain a firm foothold in the territory.

Worker bees on top of boxes of queens readied for shipment.

Without bees, the crops in this almond orchard will not be pollinated, yet bee fear is more prevalent than ever and has contributed to decreasing numbers of commercial beekeepers.

The implications of all this are many. First, within their range, the Africanized bees have largely taken over the wild hives. Professional beekeepers can keep Africanized bees—which look almost exactly like regular bees to the untrained eye—out of commercial hives. But the stronger genetics of the Africanized bees give them a serious advantage in the millions of acres where wild bees fly. For example, an online site bulletin for the U.S. National Park Service in 2008 notes that "hundreds of bee colonies call Saguaro National Park, on the Arizona's Mexican border, home. All of these colonies are now considered to be Africanized."

The next factor was bee fear. Once the phrase "killer bees" ended up on everybody's lips, bees were no longer left to just be bees. Hives, once benevolent punctuation marks in the landscape, were now something to fear. Towns began to ban hives, commercial beekeepers found it harder and harder to find public lands where they could put hives for wild blossoms, and anyone seeing a bee—much less a swarm of bees—immediately panicked and called in the pest control. Whereas once a wild swarm was something to be treasured, a kind of living wealth, now it was something to pour pesticide on until nothing moves at all. While the number of firms specializing in the removal of unwanted hives has boomed, the number of commercial beekeepers has plummeted. And as the media continues to be hysterical, the way media so enjoys being, bees have become more and more maligned: their industrious hum, as musical as a Vivaldi sonata, is now seen by all too many people to be as threatening as a growling dog.

But without the bees, the crops do not get pollinated. And a third of our daily diet simply disappears, everything from strawberries to sunflowers.

EIGHT:
The Danger

BEES, AGAIN, ARE MARVELS of evolution, a species of insect that has, from an ecological standpoint, nearly conquered the world. If Darwinism is truly survival of the fittest, few creatures have proven themselves more fit than *Apis mellifera*, which has spread out from its original homeland around the Mediterranean to nearly every corner of the globe.

That doesn't mean that the bees haven't faced dangers. One of those dangers, of course, is what this book is all about: the way we take the bees' inadvertent gift, the way honey is a delight to our tongues and souls. But how many bees have died in that harvest? How many colonies destroyed?

Humans are not the only animals who like honey. A. A. Milne's Winnie the Pooh made himself pretty famous as someone always ready to go straight for the product, whether it was by disguising himself as a cloud or simply by sitting out on a branch as the flood waters rose, counting his jars of honey to make sure he had enough to get him through the storm. Pooh was not being at all atypical; he was just being a bear, and bears love honey. Wildlife biologists who study this kind of thing are pretty sure that, just like in humans, some bears have an even bigger sweet tooth than others, and so are all the more likely to disturb the bees.

Bears can't be particularly subtle about their raids. Ideally, they find a hive, get a paw in, and get back out before the bees are too

Beekeepers at work in Idaho, their hives fenced off to protect the honey from bears.

mad. But, protected by their thick fur coats, they'll also knock hives over, rip them apart, or, according to some old stories, hold the hives underwater until the bees drown and leave only the honey behind.

Our own closest relatives, the chimps and other primates, are honey fanatics; besides, the protein in the brood cells adds a nice bit of nourishment, the way we might eat pears soaked in honey and call it a serving of fruit. Jane Goodall observed that some chimps peel bark off a stick to get to the clean interior, then use it to fish honey from the hive, licking the sweetness off like a kid with an ice cream cone. In India, tigers have gone after hives, and a host of smaller animals—most notably, the aptly named honey badger—have no hesitation about stealing from the bees.

Of course, the mere fact of being insects puts bees on the diet of a number of species of bird. The honeyguide was so named because it's easy to follow if one is looking for bees and honey. The honeyguide might also be the only animal that can easily digest beeswax, a side dish to the larvae it really wants to be eating.

An entire family of birds is known as bee-eaters, with more than twenty species. In Africa, these are startlingly beautiful creatures: the swallow-tailed bee-eater is an iridescent blue and green, the carmine bee-eater a shade of red that looks like the first blush of a rose. They all have long, thin, arched bills, and quite detailed behavior: they catch the bee (or wasp, or hornet—they aren't all that picky) on the wing, then bang it against something to kill it. Once it's dead, they pull out the stinger and squeeze the venom sacs dry before chowing down.

But more than any of those dangers for the bees, there are the dangers we ourselves pose to them, not just in the overreaction of bee fear. More desperately, we put bees at risk with the very crops

Flowers and pollen from southern California.

upon which they forage for pollen, that delicate work of plant fertilization we have enlisted them to do for us. All those big, shiny pieces of fruit in the grocery store get that way because the fields have been nearly drowned in pesticides, fertilizers, and a chemical soup that seems straight out of the imagination of a ten-year-old science geek playing in a toxic waste factory.

Bees that don't simply die outright from the poisons—from 1964 to 1971, for example, the number of colonies in Arizona was halved as billions of bees died from chemicals sprayed on cotton crops, and over roughly the same period, California lost more than 62,000 colonies per year—can be seriously harmed with long-range effects that will haunt the hive for generations. Whereas a goodly proportion of the bees probably never make it back to the hive after intense exposure to pesticides, those that do bring

OVERLEAF: *Beekeepers inspect their hives.*

Brood cells in the hive. Beekeepers divide the hive into two areas: brood cells, where eggs are laid and the bees are raised, and honey storage.

the poison back with them. That weakens the entire colony, making it much more susceptible to disease.

Beekeepers watch for unusual numbers of dead bees outside the hive (a hundred or so a day is normal in a healthy hive; more, a sign of problems). Some pesticides cause specific behavior in bees—for example, the way carbaryl or dieldrin make the bees act as if they were chilled, moving slowly, crawling around in front of the hive. Poisoned queens abandon the standard patterns they've always laid eggs in, and begin to simply lay at random. And just a few poisoned bees in the hive can completely disrupt the queen, because their presence agitates the other workers, and the business of the hive does not get done. Unattended, the queen weakens, putting the entire colony at risk.

Preventative measures can be taken, including not spraying plants when they're in bloom, spraying only at night when the bees are in the hive, or, of course, simply keeping the bees away from affected areas.

But how easy is that when one study conducted in the Pacific Northwest listed several hundred common pesticides that were harmful to bees?

If that didn't give the bees enough problems, nature herself has thrown more than a few bad things in the mix. A hundred years ago, beekeepers would have told you the largest threat to their hives was foulbrood, a spore disease that attacks larvae less than a day old. The spores would germinate in the larva's gut, beginning to grow and feed on the young bee, ultimately killing it. However, by the time that happens, the dead larvae may contain up to 100 million spores, all ready to go into the hive and infect more bees; and the spores can survive for forty years or more.

Foulbrood remains, and it is pervasive, the two strains of it— European and the more serious American—existing to some extent in nearly every hive. Chemical treatments have been developed, but for a severe infestation, the only truly effective treatment is to burn the hive. The instructions on the 2008 Queensland, Australia government Web site are almost biblical:

> Seal the hives and kill the bees. . . . Depending on how much equipment is to be burnt, a large hole is dug and a fire started. Combs burn more quickly if thrown on individually. Burn the metal lids last. Ensure all the dead bees are put into the fire otherwise some infective material may remain. Shovel any remaining material into the fire and cover the ashes with 300 mm of soil.

Foulbrood is largely under control, but new, even worse threats have arisen. For the past couple decades, scientists and beekeepers alike have been in a race to find a cure for the tracheal and varroa mites that infest bees, a dire threat to hives around the world.

Tracheal mites were first discovered around 1984. As the name implies, they move into the trachea of adult bees, ideally when the bee is around four days old, and to the most devastating effect in winter and spring. The mites set up housekeeping, lay eggs, and, like cholesterol in an artery, simply close off the bee's breathing apparatus. They also help spread other diseases and weaken the bees. Infested hives produce considerably less honey than healthy hives.

According to the 2008 Web site of the bee program at University of Georgia, Athens, infested bees will leave the colony and crawl on the grass surrounding the hive or the hive itself, but are unable to fly. In later stages, they will leave the hive altogether.

Whereas tracheal mites live inside the bees, varroa mites live outside, and are actually visible to the naked eye, at least to people who are looking really, really closely at bees. Think of them as something like a tick. They get into the brood cells of the larvae, just before the cells are capped; inside, the mite lays five or six eggs, which, when they hatch, begin to feed on the pupae.

The varroa mite, technically called *Varroa destructor*, first developed on Asian honey bees, which seemed more able to take the infestation. *Apis mellifera*, though, couldn't handle them as well when the mites began to spread. Newly hatched bees that are suffering from the mites are likely to be smaller than healthy bees, and their wings may be deformed. The University of Georgia Web site describes the symptoms: "rapid colony decline, reduced

An active, healthy colony, moments after the beekeeper has removed the top of the hive for inspection.

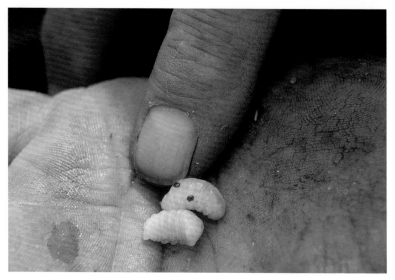

Mites on larvae. Bees are under attack from two species of mites, which have devastated hive populations around the world.

adult bee population, evacuation of the hive by crawling bees, queen supersedure, spotty brood, and abnormal brood."

So far, no cure has been found for the tracheal or varroa mites, which have spread to nearly every corner of the country—a few islands in Hawaii remain unaffected, but since the mites have shown up on Oahu, the days may be numbered for even remote colonies.

But all of that—the mites, the pesticides, the diseases—have not been as frightening to the future of bees as what has just begun. In 2006—although the earliest signs go back to 2004—a new threat to bees everywhere emerged, one that has completely changed the beekeeping landscape: colony collapse disorder, or CCD. In affected hives, the worker bees simply disappear, leaving behind perhaps a queen, a few attendants, and maybe a drone or two, sole survivors in a hive that remains full of honey. An ill or dying bee's impulse is to leave the hive, to save her sisters the trouble of removing the body or risking disease. Here, entire colonies have simply set off, as if they knew something about the health of the hives that scientists have not yet been able to figure out.

In the hardest hit areas, as many as 90 percent of the bee colonies were affected. By the end of the season, a beekeeper who had lost only half his hives was considered lucky.

At first, rampant speculation marked the discussion of what was happening: everything from global warming to pesticides to signals from cell phones was blamed. Researchers studying the problem, though, quickly came to believe that a virus was at fault, one similar or related to the Israeli Acute Paralysis Virus. IAPV is transmitted by the varroa mite, and is associated strongly with colony collapse disorder.

That is not to say though, that IAPV is causing colony collapse; merely that they seem to go together. A U.S. Department of Agriculture/Agricultural Research Service report by Kim Kaplan states that "the only pathogen found in almost all samples from honey bee colonies with CCD, but not in non-CCD colonies, was the Israeli acute paralysis virus (IAPV), a dicistrovirus that can be transmitted by the varroa mite. It was found in 96.1 percent of the CCD-bee samples."

A few other things seem to be clear: larger groups of hives are hit harder than smaller ones. Commercial beekeepers have been devastated, whereas hobby and small-scale beekeepers are still seeing their hives stay relatively safe.

Colony collapse disorder has appeared so suddenly that the logical thing is to suspect a sudden change in the landscape, whether it's cell phone towers or pesticides or viruses. But what if it's really just a threshold crossed?

Some anecdotal evidence suggests that the actual hives we use are the problem: over the years, the cells have grown, nearly forcing the bees to grow with them. But bees have a very fine sense of architecture and geography. Maybe we've just asked them to grow in ways they can't.

Which brings us to a final idea. One that may or may not be correct—maybe it's implying too much in the work of bees—but it's a question that has to be asked.

Bees have become mobile creatures in ways nature perhaps never intended. Although certainly as long ago as the ancient Egyptians, beekeepers moved hives from place to place to take advantage of crops coming into season—the Egyptians used barges on the Nile, simply bringing the boat near the fields and docking it for a few weeks, leaving the hives aboard—today, modern agriculture depends upon trucks taking bees thousands of miles from their home fields—if, indeed, they can even be said to have home fields anymore.

And it's not simply that we take bees from place to place; after all, the honey bee was an import to the Americas to begin with. But now queen bees are most commonly raised in Australia and Hawaii, shipped around the world to populate hives. Queen rearing has allowed a century of unprecedented health and control in beekeeping, the possibility of easily putting colonies into new areas, massive convenience for the keepers, and, yes, genetic control that has kept generations of colonies strong and healthy and stress-free. Raised queens can be bred for disease resistance, and a controlled environment keeps down the risks of fungus and parasites that can plague natural colonies. Colonies with greater genetic diversity are stronger, more productive, and lead to healthier swarms, which increases the strength of the colony's line and territory.

Workers in Hawaii catch queens to prepare them for shipment.

On the downside, early evidence is showing that IAPV, a virus that seems to be the trigger for colony collapse disorder, probably came to the country through imported Australian queen bees, bringing a disease to which there was no local immunity.

The current hypermobility of bees does raise issues with a species meant to thrive within a few miles of where they're born. With jets carrying people and honey and pollen to the most remote corners of the globe, a strictly local problem can become global all too quickly. The Web site for Pitcairn Island, for example, one of the most remote settlements on earth, begs people not to bring any bees or bee products to the island, where, so far, the local hives have remained disease-free.

Or maybe it's all much simpler than that. Each time the bees are moved, the hive is stripped nearly bare to make room for the new honey. And maybe that's just too much stress. Maybe all these years, bees have gone along with us, figured out ways to adapt and survive, because that's what they do, they find a way to make the

Almond orchard in California. Commercial beekeepers may truck their bees thousands of miles over the course of a year, keeping up with whatever crops are in bloom.

landscape work. But maybe now we've cost them their sense of place. Maybe we've asked bees from too far away to understand the landscape for us.

Maybe, threatened with mites and viruses, with pesticides and paranoia, it all simply comes down to the proverbial final straw.

Could it be that, like local farming movements around the world, we need to reconsider our bees as well? Surely a bee who grows and lives its entire life, the collective memory of the hive, under a particular slant of sky, in a certain scent of leaves and soil mulch, can thrive in a locale in a way that bees imported from far away, trucked on the bad-smelling interstate, loaded and unloaded by forklifts, can never compare with.

But can we thrive without that transport of bees? Modern agriculture depends on it, and without commercial beekeepers working long hours to move their hives, every field in the country would be in a world of trouble.

Whatever the cause, bees are approaching a full-blown crisis. Time and again as we worked on this book, we would go into fields that only a few years ago would have been so full of bees the plants would sound like humming refrigerators. And time and again we were surrounded by flowers, yet hardly a bee in sight.

Maybe we simply need to remember that bees have never been domesticated, that they only agree to play along with us for the sake of mutual benefit. So maybe now we should allow the bees to do what they have always done best: define their own parameters of home in the taste they stash away in the hive.

Nobody knows for sure but the bees. And so far, whatever they're saying, we're not listening closely enough.

NINE:
The Fragrant Work

"THE FIRST TIME that we open a hive there comes over us an emotion akin to that we might feel at profaning some unknown object," wrote the French poet and playwright Maurice Maeterlinck in *The Life of the Bee* (1901), "charged perhaps with dreadful surprise, as a tomb."

For a very long time after people and bees began to live more closely together, after the transition had been made between nothing but wild honey and honey brought closer to home, harvesting meant killing the hive. The only way to the sweetness was through complete destruction. Honey cells and brood cells were too closely associated.

This had wide-ranging implications. First, the beekeeper had to decide which hives to sacrifice—neither the strongest nor the weakest were good choices, as the strong would swarm and create more strong hives, and the weakest were lacking in enough honey to make a harvest worth the effort.

The reason so much destruction was necessary was a simple bit of bee psychology: when bees are home, they like tight, controlled living quarters, and they'll do all they can to close these spaces off, so that what isn't filled with comb is filled with propolis—a hard sealant that the bees make from sap or resin. This meant that in order to get at the combs, the would-be harvester

A bee smoker. Why bees mellow out in smoke is not exactly known—maybe it tranquilizes, perhaps it alarms them—but a smoker is a beekeeper's best friend when working the hive.

Bees at the hive's entrance. As part of "bee space," the architecture in which bees like to live, entrances are quite small, which allows the bees to defend the hive easily.

had to go in with all the finesse of a bear fresh out of hibernation, tearing into the hive and prying out the bees' own architecture. By the time he'd gotten to the combs themselves, the hive was a shambles, it was nearly impossible to leave enough food for the bees to survive through winter, and, as a by-product of the honey search, the brood cells, too, would all too often be damaged beyond repair.

Luckily, that all changed, thanks to a man quoted extensively in this book, L. L. Langstroth, who, in the 1850s, single-handedly saved millions of colonies, changing the millennium-old art of beekeeping. Langstroth's eureka moment came when he started really thinking about how bees like to live. Clearly, bees aren't fans of open spaces; they fill them with comb. At the same time, the bees, like inhabitants of any city, need space to get around. So, Langstroth asked, how much open space do bees need?

And so the idea of "bee space" was born. It turns out that bees are pretty specific in their desires: just $3/8$ inch between combs makes bees feel comfortable and productive.

Once he figured that out, Langstroth realized just how easy it would be to make life easier for both bees and beekeepers.

His invention couldn't have been more simple: provide the bees their happy space by arranging the hive to the bees' own specifications. Langstroth hung combs in the hive, like vertical files, leaving a careful $3/8$ inch between combs and between the combs and the side of the hive box. The hanging combs, on what's known as a form, are snug in next to each other, and are easily removable, one by one, allowing the beekeeper to check any part of the hive—or to harvest the honey without having to destroy anything.

The hundred-and-fifty-plus years since have proven that Langstroth's idea is nearly impossible to improve upon. A typical hive now consists of two or more boxes of these hanging combs: a lower one for the queen to lay eggs and propagate the hive; and upper levels, called supers, inaccessible to the queen because of a thin wire barrier, big enough for workers to pass through but not the larger queen, where the honey is stored.

Today, come harvest time, beekeepers don their protective equipment—gloves and hood, a small bellows that blows smoke onto the bees and sedates them—and go to the hives. Like the Langstroth hive itself, this equipment hasn't changed dramatically in a very long time, and what changes have occurred have merely been refinements, like a minor improvement in bellows action on the smoker.

OVERLEAF: *Forms are full and ready for harvest.*

Beekeepers have known about smoking bees for centuries, although no one has ever been entirely sure why it works. Some believe that the bees think the hive is on fire. "Instead of vainly struggling," Maeterlinck wrote, "therefore, they do what they can to safeguard the future; and, obeying a foresight that for once is in error, they fly to their reserves of honey, into which they eagerly dip in order to possess within themselves the wherewithal to start a new city, immediately and no matter where, should the ancient one be destroyed or they be compelled to forsake it." Others believe the smoke has a sedative effect, calming the bees and making them drowsy. And some just think the bees simply don't like the smoke, and so try to get away from it.

No matter. A few puffs with the bellows, and the hive is ready to be opened. The veiled beekeeper pries up the lid of the hive and peers in, ready to examine the fine work of the bees.

Depending on where the hive is geographically, it may be possible to take as many as three harvests a year. The bees themselves don't need all that much honey—forty or fifty pounds would do an average colony just fine. But their impulse is to make honey while the sun shines, to paraphrase the old proverb, and a good, productive colony can turn out 150 pounds or more of honey a year, honey that weighs down the forms in the hive.

And so at harvest time, the beekeeper simply pulls out the forms, heavy with honey cells. This is raw honeycomb, honey surrounded by beeswax—good for many ailments and maker of sweet-smelling candles.

But now, while the comb is still pregnant with honey, is its true moment. Draw a knife across the wax caps, put your tongue to the honey. The landscape has no truer gift than this, and if the hive is in your own neighborhood, that instant of the first taste will teach

The beekeeper cuts off the wax caps over the honey combs, preparing the form for extraction.

you more about the place you live than a hundred years of reading the morning paper.

The knife opens the honey cells one by one; then, the combs are loaded into an extractor—a small one is little more than frames to hold the combs, set on a spindle in a barrel—which spins the honey out through centrifugal force, the golden liquid splashing out against the extractor walls before dripping down to collection.

And then there is the overwhelming scent. Stand near an extractor, watch the drops of gold leak out, and the sense you really can't ignore is what your nose is telling you: this concentration of perfumes is the soul of flowers.

At this point in the process, there may be all kinds of impurities, up to and including bits of bees, chewy bits of wax, stray grams of propolis. But these are natural, and unlike what all too frequently happens next, don't change the taste at all.

Once the honey is out of the comb, the producer has options: the best honey is only filtered, leaving it close to its natural state, but without the stray limbs. More commercial producers cook the amber liquid, breaking it down and making it smooth and consistent, so that each jar tastes nearly the same. But you can also ferment honey—that's where mead comes from—or crystallize it, or just heat treat it, along the lines of pasteurizing milk.

The U.S. Department of Agriculture has decided that there are three *types* of honey—liquid (free from visible crystals), crystallized (just like it sounds), and partially crystallized (big surprise). They also allow two *styles* of honey, filtered and strained. And to differentiate the two, they use a simply wonderful chunk of official language which is too much fun to pass by:

> Filtered honey is honey of any type defined in these standards that has been filtered to the extent that all or most of the fine particles, pollen grains, air bubbles, or other materials normally found in suspension, have been removed.
>
> Strained honey is honey of any type defined in these standards that has been strained to the extent that most of the particles, including comb, propolis, or other defects normally found in honey, have been removed. Grains of pollen, small air bubbles, and very fine particles would not normally be removed.

Now, what does any of that taste like? They have reduced honey to an after-school special on avoiding social diseases.

Forms ready for honey extraction. The wax has been cut away and the honey-heavy combs put into the extractor, which will spin the sweet liquid out.

Jars filled with honey and comb.

Here is the problem with simply going to the grocery store and buying a plastic bear full of "honey." That honey could have come from anywhere. In fact, no matter where you live, it most likely came from either China or Brazil, and was blended, honeys from hundreds of miles apart getting poured into the same vat.

So what you taste is not the intensity of a single spot on earth, but the taste equivalent of the Tower of Babel, voices working against each other, no clear, single note appearing.

Compare this to the richness of local flavors. When you open that jar of honey, do you want to smell the place where you live, or simply some muddle? How can you truly taste honey if you're simply tasting a mud of flavors? And why would you want to, when so much real taste is so easily available?

The beekeeper, like farmers everywhere, is proud of the produce, going to great care, watching over his insect livestock, spending endless hours of labor getting the honey ready for market. And to make their honey, the bees have endlessly sampled the landscape, pulling the best moments from growing flowers and the warmth of the sun. For us to do any less, to put in any less care when we are choosing the honey for our table, would simply be ungrateful.

TEN:
Beyond Honey

HONEY OFFERS MORE than its amazing sweetness. Forget the flavor: making local honey a regular part of your diet can ease allergy symptoms; it's good for ulcers; and it's full of antioxidants, important for heart health. Honey can replace artificial sweeteners and refined sugar in recipes. It soothes sore throats when mixed with a little tea. And still the hive holds even more wonders.

Throughout history, beeswax has been worth considerably more than honey; early European importers brought Ethiopian wax back to their home countries, where it had thousands of uses beyond not quite holding together the wings of Icarus and Daedalus. Pliny boiled wax to use as both a cure for dysentery and a skin softener. Beeswax was demanded as war reparations and tribute, with tons of wax changing hands at a time. It was even acceptable as currency, as valuable (and more desirable) than cash money.

Beeswax is made by young worker bees, usually those in their second to third week of life, before they begin leaving the hive to forage. They secrete it from eight glands on their abdomens, where it comes out in scale-like shapes, almost completely colorless. The scales then get chewed by the worker bee, which turns them opaque; color comes from oils and propolis, and brood

Beeswax candles, blocks, and frame foundations. Beeswax is produced by young worker bees, who secrete it from eight glands on their abdomen.

comb tends to be darker than honeycomb, since it gets touched by more impurities.

Making wax is biologically expensive for the bees; like a sumo athlete getting ready for the next tournament, the wax-makers eat constantly, and that means bringing all the more food back to the hive. It takes over a thousand of the secreted scales to make a single gram of wax; once again, bees show the beauty of patience and cooperation.

In the hive, wax is the basic architectural building block. Bees use it to build the comb cells for the young, and the storage cells for pollen. Wax holds the honey safe until harvest. And wax is rare. A hive with a hundred pounds of honey may have only a pound and a half of wax. That means, for those harvesting the wax, not a single sliver of it could be wasted. As far back as Pliny, careful formulas for preparing and purifying beeswax had been devised and were passed on from one beekeeper to another.

The simplest way was to put the wax combs into water for cleaning, then dry them for three days—on the third day, experts would say, it was vital to keep them in darkness. On the fourth day, melt the wax with water covering all the combs, and then strain through fine cloth. Then melt it all again in the same water, and collect it in vessels smeared with honey.

But that's for low-grade, common wax. For the true, high-quality wax, known to the ancients as Punic wax (which probably came from Spain, rather than Carthage, as the name implies), the process was a bit more of an ordeal. First, expose the wax repeatedly to the open air, then heat it in sea water with a tincture of nitre. Of course, not any sea water will do—even when Julius Caesar was a baby in the olive groves that buzzed with bees, the Mediterranean had pollution problems—it has to be from the deepest parts of the sea. As the wax separates, skim off the whitest, called the flower, and put it into a jar with a little cold water. You're not done yet. Warm this up with yet more sea water, cool, repeat three times. Then dry it in a reed basket exposed to the sun and moon.

And what happened to all this wax? Medieval artisans used wax to create models for casting or as encaustic for paintings. Priests covered bodies with wax to preserve them, and it's entirely possible that in the earliest Olympic Games, athletes rubbed their bodies with wax before wrestling. Any Roman schoolboy learned how to write on wax tablets—instead of erasing, you just smoothed over the mistake. The poet Catullus suggested writing love letters in wax, which is a truly marvelous idea that should come back into style: imagine, instead of the cold letters of computer printout or email, getting something written in the sweet-smelling work of bees.

In parts of Europe, beeswax was legal tender, a perfectly acceptable means of paying your annual taxes. By the fifteenth century, wax work was recognized as a professional guild in England, the Wax Chandler's Company. Wax held together woodworks, was used as a lubricant, as a polish, and as a means of preserving food, making a moisture-proof casing.

Because honey has always been seen as a symbol of purity— that transformation of the world into food, so pure that Greek Orthodox churches even allowed eating honey during fasting periods—the wax that once held the honey, too, has long carried the idea that it is cleaner, more elemental than any other substance. According to ancient Greek stories, one of the oldest temples to Apollo was made by bees themselves from wax and feathers.

And best of all, candles made of beeswax burn long and clear,

Beeswax candles. For centuries, these provided the purest light humans could make; beeswax burns clean, long, and with almost no wax drip.

leaving only a bit of wax behind. They were used in the Eleusinian mysteries, and for centuries, they were the only source of light the Catholic Church used in its cathedrals, that lambent golden glow complemented by the scatter of sunlight breaking through stained glass. Even today, a candlemass, much more likely held under a synthetic glow than from true wax, reveals that cathedrals were designed for the flickering light of bees.

Eva Crane, proving what a genius she is at researching bees and their lore, quotes a manuscript of the Gwentian Code of the Ancient Laws of Wales from the 1200s: "The origin of bees is from Paradise and because of the sin of man they came thence, and God conferred His grace on them, and therefore the mass cannot be said without the wax."

One reason for this insistence upon beeswax was, perhaps, a slight misunderstanding of the workings of the beehive. Crane also quotes St. Augustine, who wrote that "among bees there is neither male nor female. . . . The wax of the candle is produced by the virgin bee from the flowers of the earth as a symbol of the Redeemer born of a Virgin Mother." Even today, the Paschal candle, brought in each Easter, is made of beeswax, the clean flame symbolizing the divine purity of Christ's teaching.

And even if the symbolism weren't reason enough to include the beeswax candle in worship, surely the simple fact of that gentle yellow light of the tint of honey would be enough, an echo of the bees' own color. A soft glow like the first sunrise you bother to notice.

Beeswax had other uses in religion as well. In pilgrimage churches across the continent, visitors would leave small wax figures of devotion, as prayers for a cure: someone suffering from a bad leg would leave a wax image of the leg, and talk to God to request a miracle through the work of bees.

And for the evening after worship, there is mead. Long before anyone thought of squashing grapes and letting them ferment for a while to make wine, odds are, honey was being turned into a sweet alcoholic drink. A little honey and water left in a warm place naturally ferments, and how easy it would have been to discover this as a happy accident. Plato's *Symposium* talks about becoming drunk with nectar in the time before wine.

But surely the Greek gods already knew this, doing their own imitation of bees, drinking nectar. The entire legend of Melissa, from whom the bees get their name, *mellifera*, carriers of sweetness, depends on nectar.

Robert Louis Stevenson summed up the appeal of honey to the inhabitants of the mythological world like this, in his poem "Heather Ale, A Galloway Legend":

From the bonny bells of heather,
They brewed a drink long-syne,
Was sweeter far than honey,
Was stronger far than wine.
They brewed it and they drank it,
And lay in blessed swound
For days and days together
In their dwellings underground.

Mead is not really a complicated thing to make, just a hard thing to make well. The basic recipe from a 1699 cookbook, compiled from a collection of notes of Sir Kenelm Digby, called *The Closet of Sir Kenelm Digby, Knight, Opened*:

> Take 18 quarts of spring-water, and one quart of honey; when the water is warm, put the honey into it. When it boileth up, skim it very well, and continue skimming it, as long as any scum will rise. Then put in one Race of Ginger (sliced in thin slices), four Cloves, and a little sprig of green Rosemary. Let these boil in the Liquor so long, till in all it have boiled one hour. Then set it to cool, till it be blood-warm, and then put to it a spoonful of Ale-yest. When it is worked up, put it into a vessel of a fit size; and after two or three days, bottle it up. You may drink it after six weeks, or two months.

After you've laid yourself out drunk on fine mead, perhaps it's time to see the doctor. Honey and other bee products are increasingly important as medical treatments well beyond the standard

Mead, one of the oldest alcoholic drinks known, and one held sacred in many cultures.

sore throat cure of tea with honey and lemon. As mentioned earlier, some believe local honey can help assuage allergies—like an inoculation, filling yourself with the ghosts of pollen.

Studies have shown that a smear of honey on a bandage can help wounds heal more quickly. It's particularly effective to prevent the growth of bacteria; one theory is that honey, although a liquid, is actually quite dry, so it acts as a sponge, sucking moisture right out of the bacteria and killing them. Another possible explanation is the acidic nature of honey working as an antiseptic. Honey-treated bandages are especially useful on the tender skin of burn victims.

Regular doses of honey can help heal ulcers and ease gastroenteritis and other digestive tract diseases; recent research even shows it might be a useful treatment for MRSA, a staph infection that has been running rampant the past few years.

Bee stings themselves can be life-threatening to some people, causing anaphylactic shock, a swelling that, in severe cases, can close down the windpipe and zero out blood pressure. But the stings, some healers are finding, can also be helpful for arthritis, rheumatism, back pain, and multiple sclerosis. Some have even had success using bee stings to treat chronic fatigue syndrome.

The ideal treatment is to use a live bee to sting a point—like an acupuncture meridian. The bees' venom is most potent in the late spring and summer, when they are well-fed on fresh honey. If no live bees are available, one can actually buy bee venom, which an apitherapist will inject through more traditional medical methods. And, unlike having the live bee sting, collected venom does not harm the bees—at least not much. A mild electrical charge stimulates the bees into stinging the collector sheet. The stinging motion produces the venom, but there is nothing for the bee's stinger

Scrapings from the hive include propolis, a hard substance that bees use for structure and humans use for a variety of medicinal products.

to penetrate, and so, instead of the barbs eviscerating the bee when she pulls the stinger back, the stinger slides harmlessly off the plate and the bee lives to sting again. Meanwhile, the venom itself dries and can be scraped off for later treatment. A single sting carries the equivalent of about 0.1 mg of dried venom.

And we're still not done with the possibilities. Bees use propolis as architectural mortar, proving that they are the alchemists of the insect world. Propolis, which tends to be dark in color, is made from the resin of trees—in temperate forests, poplars and the sticky sap of conifers are favorites, but in other regions, the bees may use resin from flowers, or even artificial sources. The bees use this substance to chink cracks, close up spaces that violate their $^3/_8$-inch rule, reinforce the hive, and perhaps even help control the hive's acoustics. Dried propolis is hard and tough, impervious to weather; in an ambitious hive, a hammer and chisel might not be enough to dislodge all the propolis.

But once it is dislodged, humans have found uses for it the bees

might not have considered. Propolis is used in treatments for inflammation, ulcers, and burns. Some believe it may have salutary effects on the heart, and reduce cataracts. And, by the same logic that suggests local honey may be useful in treating allergies, propolis, too, has been used as an antihistamine. As with most bee products, propolis is sterile, and may have both antibiotic and antifungal attributes. It also helps reduce canker sores. If all that weren't enough, the Italian violin maker Stradivari used propolis as part of the varnish that is considered the key to the Stradivarius sound.

Yet more magic goes on in the hive. When a queen egg is laid, it is no different than a worker egg. What makes the change is royal jelly, another substance humans are finding works for them as well as for bees.

All worker bees get some royal jelly during their larval stage; a queen-to-be, though, gets only royal jelly, which is secreted from glands in the heads of young worker bees. The exclusive diet of royal jelly triggers the development of ovaries, as well as a cascade of chemical changes that bring about the transformation of worker larvae to queen.

For medicinal use, royal jelly is collected from cells roughly four days old. It's a labor intensive project, and because the jelly can spoil easily, time-sensitive. Chemically, the jelly is fairly simple: about 66 percent water, 12.5 percent crude protein and amino acids, and roughly 11 percent simple sugars. Toss in some trace minerals, the standard antibacterial and antimicrobial qualities found throughout bee products, and it seems a straightforward product of the hive.

Beekeeper taking royal jelly from a queen cell.

Products of the beehive: honey and beeswax form the basis of many skin-care items; propolis and royal jelly are valued for their pharmaceutical uses.

But research suggests that royal jelly may help fight Parkinson's disease and perhaps even Alzheimer's, by stimulating the growth of glial and neural cells. It may have some effect at lowering cholesterol, fighting cancers, and treating inflammatory diseases. And, like most bee products, royal jelly can also speed the healing of wounds.

Need more? Bee pollen. The idea is that bees are picky about the pollen they choose, which means what makes it to the hive is the best, most nutritionally sound pollen. Pollen is great for vitamin A, for flavonoids, for cytochromes. It's good, believers say, for weight loss, sex drive, prostate health, sperm count. Pollen helps build muscles and endurance, and mental clarity. Tests are ongoing to see if it can fight cancer.

But enough of the serious side of bees and what they can do. Honey tastes and smells good, and isn't that enough? Surely it appears so for the hundreds of manufacturers of soaps and lotions that include honey in their ingredients; it's proven to smooth skin and reduce fine wrinkles.

And where would the candy market be without honey? So many recipes that today are drowning in sugar were once made with honey. A spoonful of honey is nature's original energy drink, dating thousands of years before humans stumbled upon caffeine.

Try soaking pears or figs in honey. Get in touch with your inner bear and let nuts or cherries marinate in honey until they are imbued with its flavor. Drizzle honey over goat cheese and anise to find the true meaning of bittersweet.

Or simply raise your glass of mead in a toast: to the bees.

ELEVEN:
The Taste of the World

FLOWER BY FLOWER, grain by grain, the bees continue their work, the only work they've ever known. The landscape opens to them, beckons with colors and shapes as bold to the bees as the neon signs in Times Square are to a tourist, conveying every bit as much information: here, in these white petals, is the fine, sticky sparseness of orange pollen; there, hidden behind crenellations that look like nothing so much as the hat Jughead wore on old Archie comics, is the impossible flavor of the tupelo blossom, rising miraculously from the swamp. In the afternoon sunlight, the *Disa versicolor* orchid gives off the last wafts of the day's scent, and worker bees flock to its bright, deep red petals, a color they can't even see, but know how to turn into grace.

Across the fields, in the nooks and crannies of the earth, across fields and plains, on mountainsides that shelter only single flowers clinging to bare rock, bees pause, they gather, they eat, and then they fly, with perfect, unerring accuracy, back to the hive, where the magical transformation of honey begins.

Now, here's a simple fact of nature, something we cannot afford to forget: roughly three-fifths of the plants in our country, and a full one-third of our daily diet, depend on bee pollination, that little by-product in the creation of honey. No bees means there is no onion, no kiwifruit, no okra. Without bees, we have no

A close-up of a bee on an almond flower. Bees are responsible for pollinating roughly one-third of the world's food crops.

celery, no mustard, no rapeseed, no cabbage or cauliflower. If the bees disappear, so do watermelons and tangerines. Cucumbers, cantaloupes, hazelnuts, gone. Soybeans, coconuts, sunflowers. Avocados, apricots, and cherries. Pomegranates.

Apples. Without bees and their patient labor, we do not even have apples.

So we always must be grateful for this miracle, for the fact that, even under threat from ever-encroaching civilization, from disease and frost and drought, the bees still keep doing what they do, making us a gift of the landscape.

Actually, if you really stop to think about it, they make the landscape itself possible.

Winnie the Pooh once said, with only a bit of bear-ish bias, "the only reason for being a bee that I know of is making honey. . . . And the only reason for making honey is so as *I* can eat it."

But the return is so much greater than that. A jar of local honey is potential made actual, the possibility of understanding where one lives or travels in the most vivid way possible, on the tongue. Oh, we may all too often face over-processed honey as generic as a soft drink, but when you know there is more, it's like walking through a door of what you thought was a tiny room and finding a new universe on the other side. It's like the first morning when you realize you're in love.

Consider this: a summer day, the sky wide and impossibly blue, blossoms waving a hundred colors everywhere they can find just a few grains of dirt to grow in. Maybe most people walk by, never noticing the blooms or the small visitors, six legs, yellow body, wings that are surely too fragile to allow for flight.

Some of the fruits and vegetables that depend on bees for pollination.

But that visitor, the bee, one of perhaps 30,000 or so from her hive—just another hive among the millions that enrich our globe—out foraging that day, pauses in the flower for a moment. The bee walks the petals like a farmer checking a field as familiar as childhood. Then she dips into the flower; pollen clings to her legs while she takes her fill of nectar.

Then, without a pause, without a moment's rest, without a conscious thought to the creation of such staggering beauty, she's off, looking for the next flower, navigating the world through its most fragile landmarks, turning her every moment into the language of honey.

The bee takes to the air again, carrying her on a route that is her share of what a gallon of honey requires to gather: a journey of such great distance that in sheer miles that gallon of flavor may as well require a trip to the moon and back. A journey performed by insects not much larger than a fingernail, flying on wings so delicate they can wear out from the friction of the air itself, as they beat more than 200 times per minute.

Not concerned with us at all, not even considering us, the bee, her legs heavy with pollen, flies on, merely doing the same thing bees have always done: the fine, delicate work of perfect harmony, bee and bee, bee and flower, bee and the spinning globe.

But let's say this bee, heading back to the hive, flies by someone. Let's say the bee flies past a beautiful woman who is walking out of a shop, smile on her face. In the woman's hand, a small jar, full of liquid the color of the landscape after a storm. It has come to her through a simple transaction, but one unlike any other purchase she'll make that day. For in her hand is something that tastes like sunshine and cloudy days.

Honey comes in endless varieties and flavors.

In her hand is something as miraculous as her smile itself: the taste of a pure and flawless moment in the world. In her hand is a jar of thick honey, a light held still and liquid, the tangible fossil of fields of flowers in days of sunshine, afternoons of a heat that feels like a comfort, and the first fat drops of a summer rainstorm; the record of soil and chemical and color and growing season, of the slant of the moon and the shadow of a perching bird.

In her hand is a record of the world.

In her hand, in that jar of honey, is the flavor that haunts forever, that drop whose first touch on your tongue you can never forget, the perfect taste of sweetness.

Bibliography

Aristotle. 2002. *History of Animals*. Translated by A. L. Peck. Boston: Loeb Classical Library. First published 343 BC.

Budge, E. A. Wallis. 1989. *The Mummy: A Handbook of Egyptian Funerary Archaelogy*. NY: Dover Publications. First published 1893 by University Press, Ltd., London.

Carter, Howard. 2003. *The Tomb of Tut-ankh-amen*. London: Duckworth Publishers. First published 1927 by Cassell and Company, London.

Chinmayananda, Swami. 1980. *Thousand Ways to the Transcendental: Vishnu Sahasranama*. India: Central Chinmayananda Mission Trust.

Columella, Lucius Junius Moderatus. 1941. *De Re Rustica*. Translated by H. B. Ash. Boston: Loeb Classical Library. First published 68– 42 AD.

Crane, Eva. 1999. *The World History of Beekeeping and Honey Hunting*. NY: Routledge.

Digby, Kenelm. 2005. *The Closet of Sir Kenelm Digby Knight Opened*. Edited by Anne MacDonell. London: Philip Lee Warner. First published 1669.

Gamble, Reginald. 1943. *Bee Keeping: Young Farmers' Club*, Booklet No. 2, fourth edition. England: National Federation of Young Farmers.

Grimm, Jacob and Wilhelm. 1976 ed. *The Complete Grimm's Fairy Tales*. New York: Pantheon. First published 1812.

Hopkins, Isaac. 1866. "Method of Queen Rearing." *The Australasian Bee Manual*.

Kaplan, Kim. 2007. "Genetic Survey Finds Association Between CCD and Virus." http://www.ars.usda.gov/IS/pr/2007/070906. htm (accessed August 2008).

Langstroth, L. L. 2004. *Hive and the Honey-Bee: The Classic Beekeeper's Manual*. New York: Dover Publications. First published 1853 by Hopkins, Bridgman, & Co., Northhampton, MA.

Longfellow, Henry Wadsworth. 2006. *Song of Hiawatha*. New York: Dover Publications. First published 1855 by Ticknor and Fields, Boston.

Maeterlinck, Maurice. 2003. *The Life of the Bee*. Translated by Alfred Sutro. Available on Project Gutenberg: http://www.gutenberg. org/etext/4511. First published 1901 by Dodd, Mead and Company, New York.

Pripovedi S Panjev. Radovljica: Cebelarski Muzej. Unpaginated museum brochure.

Queensland Government, Department of Primary Industries and Fisheries. "Quarantine and Management Plan." http://www.dpi. qld.gov.au/cps/rde/dpi/hs.xsl/27_10636_ENA_HTML. htm#Quarantine_and_management_plan (accessed August 2008).

Ransome, Hilda M. 2004. *The Sacred Bee in Ancient Times and Folklore*. NY: Dover Publications. First published 1937 by George Allen & Unwin, London.

Shakespeare, William. 1998. *Henry V*. Edited by Gary Taylor. NY: Oxford University Press. First published 1623.

Shakespeare, William. 2003. *The Tempest*. Edited by Peter Hulme and William H. Sherman. NY: W. W. Norton. First published 1623.

Sophocles, *Oedipus Rex*. 2005. Edited by Elizabeth Osborne; translated by J.E. Thomas. DE: Prestwick House, Inc. First published 429 BC.

United States Department of Agriculture, Agriculture Marketing Service. 1985. "United States Standards for Grades of Extracted Honey." http://www.ams.usda.gov/AMSv1.0/getfile?dDocName =STELDEV3011895 (accessed August 2008).

University of California, Davis. 2008 online catalog. "Art of Queen Rearing Workshop." http://entomology.ucdavis.edu/courses/ beeclasses/queenrearing.html (accessed August 2008).

University of Georgia, Athens. "Honey Bee Disorders: Tracheal Mites." http://www.ent.uga.edu/bees/Disorders/Tracheal_ mites.htm (accessed August 2008).

———. "Honey Bee Disorders: Varroa Mites." http://www.ent.uga. edu/bees/Disorders/Varroa_mites.htm (accessed August 2008).

U.S. National Park Service. "Plan Your Visit" site brochure. www.nps.gov/sagu/planyourvisit/upload/Africanized%20Honey %20Bees.pdf (accessed August 2008).

Virgil. 1953. *The Georgics*. Translated by John Dryden (first translation 1697). NY: Heritage Press. First published 29 BC.

Von Frisch, Karl. 1967. *The Dance Language and Orientation of Bees*. Boston: The Belknap Press of Harvard University Press.

Wildman, Thomas. 1770. *Treatise on the Management of Bees*, London: Strahan and Cadell.

William the Conqueror. 2004. *The Domesday Book*. Translated by F. Martin and Ann Williams. New York: Penguin Classics. First published 1086.